高等职业教育"十三五"精品规划教材

计算机专业英语
（第三版）

主 编 孙建忠 白凤仙

中国水利水电出版社
www.waterpub.com.cn
·北京·

内 容 提 要

本书以计算机和网络技术为背景，对计算机英语进行了全面的描述。全书涉及计算机的历史与发展、计算机的组成结构、数据结构、操作系统、软件工程、计算机语言、因特网、万维网、网络与安全、物联网、云计算、数据库管理系统、多媒体和动画等 14 个主题，并介绍计算机技术与应用的一些最新发展。

本书所选材料语言规范、内容新颖、完整实用。每一章均包括：学习指导与小结、课文、注释与译文、阅读材料、练习与专业英语专题等内容。专业英语专题包括计算机词汇及其构成规律、科技英语的阅读与翻译技巧、科技论文及其摘要的写作、商业书信以及日常求职英语等必备知识。

本书可作为高职高专计算机专业的专业英语教材，也可作为使用计算机的广大科技工作者的参考书。

本书配有电子教案，读者可以从中国水利水电出版社网站和万水书苑免费下载，网址为：http://www.waterpub.com.cn/softdown 和 http://www.wsbookshow.com。

图书在版编目（CIP）数据

计算机专业英语 / 孙建忠，白凤仙主编. -- 3版
. -- 北京：中国水利水电出版社，2016.9（2019.12重印）
  高等职业教育"十三五"精品规划教材
  ISBN 978-7-5170-4660-8

Ⅰ. ①计… Ⅱ. ①孙… ②白… Ⅲ. ①电子计算机－英语－高等职业教育－教材 Ⅳ. ①TP3

中国版本图书馆CIP数据核字(2016)第207785号

责任编辑：杨庆川　　　加工编辑：庄　晨　　　封面设计：李　佳

书　名	高等职业教育"十三五"精品规划教材 计算机专业英语（第三版）JISUANJI ZHUANYE YINGYU
作　者	主　编　孙建忠　白凤仙
出版发行	中国水利水电出版社 （北京市海淀区玉渊潭南路1号D座　100038） 网址：www.waterpub.com.cn E-mail: mchannel@263.net（万水） 　　　　sales@waterpub.com.cn 电话：（010）68367658（营销中心）、82562819（万水）
经　售	全国各地新华书店和相关出版物销售网点
排　版	北京万水电子信息有限公司
印　刷	三河市鑫金马印装有限公司
规　格	184mm×260mm　16开本　14.5印张　361千字
版　次	2005年4月第1版　2005年4月第1次印刷 2016年9月第3版　2019年12月第2次印刷
印　数	4001—6000册
定　价	28.00元

凡购买我社图书，如有缺页、倒页、脱页的，本社营销中心负责调换

版权所有·侵权必究

第三版前言

今天，人类已经进入了"互联网+"时代，物联网，云计算，大数据，人类正在迎来又一次信息产业浪潮。地球村的出现打破了传统的时空观念，使人们与外界乃至整个世界的联系更为紧密。由于诸多原因，国际上最通用的语言还是英语；而因特网的普及，更巩固了英语作为跨文化交往通用语言的地位。因此，为了掌握最新的信息技术，了解信息技术的发展动向，必须具备较高的英语水平。本书的编写目的，首先是让学生掌握计算机专业英语的基本术语，了解一些计算机专业的基本知识；其次是为了介绍信息技术的一些最新发展。

本书以计算机与网络技术为背景，充分考虑了计算机英语的复杂性和新颖性，针对高职高专的教学特点，精心组织，合理选材。主要内容包括：计算机的历史与发展、计算机的组成结构、数据结构、操作系统、软件工程、计算机语言、因特网、万维网、网络与安全、数据库管理系统、多媒体、动画、物联网和云计算等的基本概念等共14章。

本书为第3版，第1、2版受到了广大读者的关心和支持，热心的读者还专为第1版课文提供了部分译文，借本书第3版出版的机会对广大读者的支持再次表示衷心的感谢。

根据教材的使用情况及计算机技术的发展，对本书进行了修订。在延续第1、2版编写风格的基础上，对教材内容进行了更新，删去了一些过时的内容，增加了反映计算机技术和信息技术发展水平的新内容；全书按照计算机基础、计算机网络与计算机应用三个部分进行了组织；考虑到学生的实际需要，系统而扼要地介绍了计算机词汇构成规律、科技英语的阅读与翻译技巧、科技论文及其摘要的写作以及求职英语和广告英语的特点；为了便于教学，还附有参考译文。

本书由大连理工大学孙建忠、白凤仙主编，主要编写人员具体分工为：第1、2、3、13、14章由孙建忠编写，第4、5、6、7章由程立编写，第8、9、10、11、12章由白凤仙和姚卫红编写。参加本书编写的还有李若芬、周龙、秦世宏、李智、李梅、刘翌南、李琳等。

由于作者水平有限，编写时间仓促，本书在编写过程中难免出现疏漏，恳请读者不吝赐教，邮件地址：sjzbfx@163.com。

编　者
2016年8月

目 录

第三版前言

Chapter 1　The History and Future of Computers ……………………………… 1
学习指导 ……………………………………… 1
1.1　The Invention of the Computer ……… 1
　1.1.1　The ENIAC ……………………… 1
　1.1.2　The UNIVAC I …………………… 2
1.2　Computer Generations ………………… 2
　1.2.1　First-Generation Computers: 1951~1958 … 3
　1.2.2　Second-Generation Computers: 1959~1963 ……………………… 3
　1.2.3　Third-Generation Computers: 1964~1970 ……………………… 4
　1.2.4　Fourth-Generation Computers: 1971~? ………………………… 4
　1.2.5　Generationless Computers ………… 5
Reading Material: Classes of Computing Applications and Their Characteristics …… 6
科技英语的特点 …………………………… 7
Exercises …………………………………… 10

Chapter 2　Basic Organization of Computers … 11
学习指导 …………………………………… 11
2.1　Introduction …………………………… 11
2.2　System Buses ………………………… 12
2.3　Instruction Cycle ……………………… 13
2.4　CPU ORGANIZATION ……………… 15
Reading Material: Eight Great Ideas in Computer Architecture ……………………… 17
计算机英语专业词汇的构成 ……………… 19
Exercises …………………………………… 23

Chapter 3　Binary System and Boolean Algebra ………………………………… 24
学习指导 …………………………………… 24
3.1　The Decimal System ………………… 24

3.2　The Binary System …………………… 26
3.3　Boolean Algebra ……………………… 26
数学公式的读法（Pronunciation of mathematical expressions）……………………… 28
Exercises …………………………………… 31

Chapter 4　Elementary Data Structures ……… 32
学习指导 …………………………………… 32
4.1　Stacks and queues …………………… 32
　4.1.1　Stacks …………………………… 32
　4.1.2　Queues …………………………… 33
4.2　Linked lists …………………………… 35
　4.2.1　Searching a linked list …………… 36
　4.2.2　Inserting into a linked list ………… 36
　4.2.3　Deleting from a linked list ………… 36
　4.2.4　Sentinels ………………………… 37
Reading Material: Related Concepts ……… 38
常用英汉互译技巧 ………………………… 40
Exercises …………………………………… 45

Chapter 5　Operating System ………………… 46
学习指导 …………………………………… 46
5.1　OS Functions ………………………… 46
　5.1.1　Resource allocation and related functions …………………………… 46
　5.1.2　User interface related functions …… 48
5.2　FreeBSD vs. Linux vs. Windows 2000 …… 49
Reading Material: RATs …………………… 52
被动语态的译法 …………………………… 54
Exercises …………………………………… 57

Chapter 6　Software Engineering ……………… 58
学习指导 …………………………………… 58
6.1　Basic Software Concepts …………… 58
　6.1.1　Application Software …………… 59
　6.1.2　System Software ………………… 59

6.2 The Software Life Cycle ········· 60
  6.2.1 System engineering and analysis ······· 61
  6.2.2 Software requirements analysis ········ 61
  6.2.3 Design ········ 61
  6.2.4 Coding ········ 61
  6.2.5 Testing ········ 61
  6.2.6 Maintenance ········ 62
6.3 Prototyping ········ 63
Reading Material: Software Engineering Methodologies ········ 64
长定语（从句）的翻译技巧之一 ········ 66
Exercises ········ 69

Chapter 7 Programming Language ········ 70
本章学习指导 ········ 70
7.1 Introduction to Programming Language ····· 70
7.2 Object-oriented Programming ········ 72
7.3 OMG's Unified Modeling Language(UML) ··· 73
Reading Material: Programming Paradigms ······· 75
长定语（从句）的翻译技巧之二 ········ 78
Exercises ········ 81

Chapter 8 The Internet ········ 82
学习指导 ········ 82
8.1 The Internet: Key Technology Concepts ····· 82
8.2 Other Internet Protocols and Utility Programs ········ 88
8.3 Internet Service Providers ········ 89
Reading Material: Transition from IPv4 to IPv6 ··· 91
英语长句的翻译 ········ 92
Exercises ········ 96

Chapter 9 The World Wide Web ········ 97
学习指导 ········ 97
9.1 Hypertext ········ 97
9.2 Markup Languages ········ 98
9.3 Web Servers and Clients ········ 100
9.4 Web Browsers ········ 101
Reading Material: Features of The Internet and The Web ········ 102
学术论文的英文写作简介 ········ 107
Exercises ········ 109

Chapter 10 Network Security ········ 110
学习指导 ········ 110
10.1 Secure Networks and Policies ········ 110
10.2 Aspects of Security ········ 111
10.3 Responsibility and Control ········ 111
10.4 Integrity Mechanism ········ 111
10.5 Access Control and Passwords ········ 112
10.6 Encryption and Privacy ········ 112
10.7 Public Encryption ········ 113
10.8 Authentication with Digital Signatures ···· 114
10.9 Packet Filtering ········ 115
10.10 Internet Firewall Concept ········ 116
Reading Material: KINDS OF SECURITY BREACHES ········ 117
科技论文标题的写法 ········ 118
Exercises ········ 120

Chapter 11 Database System ········ 121
学习指导 ········ 121
11.1 Overview ········ 121
11.2 Database Models ········ 123
  11.2.1 Flat File ········ 123
  11.2.2 Relational ········ 123
  11.2.3 Hierarchical ········ 124
  11.2.4 Other Database Models ········ 125
11.3 Data Mining ········ 126
英文论文引言的写作技巧 ········ 127
Exercises ········ 134

Chapter 12 Multimedia and Compuer Animations ········ 135
学习指导 ········ 135
12.1 Multimedia ········ 135
  12.1.1 Visual Elements ········ 136
  12.1.2 Sound Elements ········ 136
  12.1.3 Organizational Elements ········ 137
  12.1.4 Multimedia Applications ········ 137
12.2 Computer Animation ········ 138
  12.2.1 Design of Animation Sequences ····· 138
  12.2.2 General Computer-Animation Functions ········ 139

英文摘要的写作技巧 …………………… 140
Exercises ……………………………… 144

Chapter 13　Anatomy of the Internet of Things 145
学习指导 ……………………………… 145
13.1　Traditional Internet Protocols Aren't the Solution for Much of the IoT ………… 146
　　13.1.1　Introducing the "Chirp" …………… 146
　　13.1.2　Functionality the IoT Needs—and Doesn't ……………………… 147
　　13.1.3　Efficiency Out of Redundancy …… 148
13.2　It's All Relative ……………………… 150
　　13.2.1　Format Flexibility ………………… 151
　　13.2.2　Private Markers for Customization and Extensibility ………………… 151
　　13.2.3　Addressing and "Rhythms" ……… 152
　　13.2.4　Family Types ……………………… 152
13.3　Applying Network Intelligence at Propagator Nodes ……………………… 153
　　13.3.1　Transport and Functional Architectures ……………………… 154
　　13.3.2　Functional Network Topology …… 156
　　13.3.3　Defined by Integrator Functions …… 156
　　13.3.4　Harvesting Information from the IoT … 157
　　13.3.5　Programming and "Bias" ………… 157
　　13.3.6　Receiver-Oriented Selectivity …… 158
求职英语简介 …………………………… 159
Exercises ……………………………… 162

Chapter 14　Cloud Computing ……………… 163
学习指导 ……………………………… 163
14.1　Definitions …………………………… 163
14.2　Related Technologies for Cloud Computing …………………………… 165
14.3　Cloud Service Models ……………… 168
14.4　Cloud Deployment Models ………… 169
14.5　Public Cloud Platforms: State-of-the-Art … 170
14.6　Business Benefits of Cloud Computing … 172
广告文体简介 …………………………… 173
Exercises ……………………………… 176

参考译文 ………………………………… 177
参考文献 ………………………………… 226

Chapter 1   The History and Future of Computers

学习指导

20世纪40年代，世界上诞生了第一台电子计算机。此后，随着真空管、晶体管、集成电路与超大规模集成电路的发展及其在计算机中的应用，计算机从第一代发展到第四代。而今天，由于科学技术的变化日新月异，计算机的发展进入了"无代"时代。通过本章学习，读者应掌握以下内容：

- 现代计算机的共同特征和各代计算机的特点；
- 计算机技术的发展趋势；
- 了解科技英语的特点，掌握科技英语翻译要点。

1.1   The Invention of the Computer

It is hard to say exactly when the modern computer was invented. Starting in the 1930s and through the 1940s, a number of machines were developed that were like computers. But most of these machines did not have all the characteristics that we associate with computers today. These characteristics are that the machine is electronic, that it has a stored program, and that it is general purpose.

One of the first computerlike devices was developed in Germany by Konrad Zuse in 1941. Called the Z3, it was general-purpose, stored-program machine with many electronic parts, but it had a mechanical memory. Another electromechanical computing machine was developed by Howard Aiken, with financial assistance from IBM, at Harvard University in 1943. It was called the Automatic Sequence Control Calculator Mark I, or simply the Harvard Mark I. Neither of these machines was a true computer, however, because they were not entirely electronic.

1.1.1   The ENIAC

Perhaps the most influential of the early computerlike devices was the Electronic Numerical Integrator and Computer, or ENIAC. It was developed by J. Presper Eckert and John Mauchly at the University of Pennsylvania. The project began in 1943 and was completed in 1946. The machine was huge; it weighed 30 tons and contained over 18,000 vacuum tubes.

The ENIAC was a major advancement for its time. It was the first general-purpose, electronic computing machine and was capable of performing thousands of operations per second. It was controlled, however, by switches and plugs that had to be manually set. Thus, although it was a general-purpose electronic device, it did not have a stored program. Therefore, it did not have all the characteristics of a computer.

While working on the ENIAC, Eckert and Mauchly were joined by a brilliant mathematician, John von Neuman. Together, they developed the idea of a stored program computer. This machine, called the Electronic Discrete Variable Automatic Computer, or EDVAC, was the first machine whose design included all the characteristics of a computer. It was not completed, however, until 1951.

Before the EDVAC was finished, several other machines were built that incorporated elements of the EDVAC design of Eckert, Mauchly, and von Neuman. One was the Electronic Delay Storage Automatic Computer, or EDSAC, which was developed in Cambridge, England. It first operated in May of 1949 and is probably the world's first electronic stored-program, general-purpose computer to become operational. The first computer to operate in the United States was the Binary Automatic Computer, or BINAC, which became operational in August of 1949.

1.1.2 The UNIVAC I

Like other computing pioneers before them, Eckert and Mauchly formed a company in 1947 to develop a commercial computer. The company was called the Eckert-Mauchly Computer Corporation. Their objective was to design and build the Universal Automatic Computer or UNIVAC. Because of difficulties of getting financial support, they had to sell the company to Remington Rand in 1950. Eckert and Mauchly continued to work on the UNIVAC at Remington Rand and completed it in 1951. Known as the UNIVAC I, this machine was the first commercially available computer.

The first UNIVAC I was delivered to the Census Bureau and used for the 1950 census. The second UNIVAC I was used to predict that Dwight Eisenhower would win the 1952 presidential election, less than an hour after the polls closed. The UNIVAC I began the modern of computer use.

New Words & Expressions

computerlike	a. 计算机似的	electromechanical	a. 机电的，电机的
vacuum tubes	真空管	Census Bureau	人口普查局
thousands of	成千上万的	known as	通常所说的，以……著称

Abbreviations

ENIAC (Electronic Numerical Integrator and Computer) 电子数字积分计算机，ENIAC 计算机
EDSAC (Electronic Delay Storage Automatic Computer) 延迟存储电子自动计算机
BINAC (Binary Automatic Computer) 二进制自动计算机
UNIVAC (Universal Automatic Computer) 通用自动计算机

1.2 Computer Generations

Since the UNIVAC I computers have evolved rapidly. Their evolution has been the result of changes in technology that have occurred regularly. These changes have resulted in four main generations of computers.

1.2.1 First-Generation Computers: 1951~1958

First-generation computers were characterized by the use of vacuum tubes as their principal electronic component. Vacuum tubes are bulky and produce a lot of heat, so first-generation computers were large and required extensive air conditioning to keep them cool. In addition, because vacuum tubes do not operate very fast, these computers were relatively slow.

The UNIVAC I was the first commercial computer in this generation. As noted earlier, it was used in the Census Bureau in 1951. It was also the first computer to be used in a business application. In 1954, General Electric took delivery of a UNIVAC I and used it for some of its business data processing.

The UNIVAC I was not the most popular first-generation computer, however. This honor goes to the IBM 650. It was first delivered in 1955 before Remington Rand could come out with a successor to the UNIVAC I. With the IBM 650, IBM captured the majority of the computer market, a position it still holds today.

At the same time that hardware was evolving, software was developing. The first computers were programmed in machine language, but during the first computer generation, the idea of programming language translation and high-level languages occurred. Much of the credit for these ideas goes to Grace Hopper, who, as a Navy lieutenant in 1945, learned to program the Harvard Mark I. In 1952, she developed the first programming language translator, followed by others in later years. She also developed a language called Flow-matic in 1957, which formed the basis for COBOL, the most commonly used business programming language today.

Other software developments during the first computer generation include the design of the FORTRAN programming language in 1957. This language became the first widely used high-level language. Also, the first simple operating systems became available with first-generation computers.

1.2.2 Second-Generation Computers: 1959~1963

In the second generation of computers, transistors replaced vacuum tubes. Although invented in 1948, the first all-transistor computer did not become available until 1959. Transistors are smaller and less expensive than vacuum tubes, and they operate faster and produce less heat. Hence, with second-generation computers, the size and cost of computers decreased, their speed increased, and their air-conditioning needs were reduced.

Many companies that had not previously sold computer entered the industry with the second generation. One of these companies that still makes computers is Control Data Corporation (CDC). They were noted for making high-speed computers for scientific work.

Remintong Rand, now called Sperr-Rand Corporation, made several second-generation UNIVAC computers. IBM, however, continued to dominate the industry. One of the most popular second-generation computers was the IBM 1401, which was a medium-sized computer used by many businesses.

All computers at this time were mainframe computers costing over a million dollars. The first minicomputer became available in 1960 and cost about $120,000. This was the PDP-1, manufactured by Digital Equipment Corporation (DEC).

Software also continued to develop during this time. Many new programming languages were designed, including COBOL in 1960. More and more businesses and organizations were beginning to use computers for their data processing needs.

1.2.3 Third–Generation Computers: 1964~1970

The technical development that marks the third generation of computers is the use of integrated circuits or ICs in computers. An integrated circuit is a piece of silicon (a chip) containing numerous transistors. One IC replaces many transistors in a computer; result in a continuation of the trends begun in the second generation. These trends include reduced size, reduced cost, increased speed, and reduced need for air conditioning.

Although integrated circuits were invented in 1958, the first computers to make extensive use of them were not available until 1964. In that year, IBM introduced a line of mainframe computers called the System/360. The computers in this line became the most widely used third-generation machines. There were many models in the System/360 line, ranging from small, relatively slow, and inexpensive ones, to large, very fast, and costly models. All models, however, were compatible so that programs written for one model could be used on another. This feature of compatibility across many computers in a line was adopted by other manufacturers of third-generation computers.

The third computer generation was also the time when minicomputers became widespread. The most popular model was the PDP-8, manufactured by DEC. Other companies, including Data General Corporation and Hewlett-Packard Company, introduced minicomputers during the third generation.

The principal software development during the third computer generation was the increased sophistication of operating systems. Although simple operating systems were developed for first-and second-generation computers, many of the features of modern operating systems first appeared during the third generation. These include multiprogramming, virtual memory, and time-sharing. The first operating systems were mainly batch systems, but during the third generation, interactive systems, especially on minicomputers, became common. The BASIC programming language was designed in 1964 and became popular during the third computer generation because of its interactive nature.

1.2.4 Fourth–Generation Computers: 1971 ~ ?

The fourth generation of computers is more difficult to define than the other three generations. This generation is characterized by more and more transistors being contained on a silicon chip. First there was Large Scale Integration (LSI), with hundreds and thousands of transistors per chip, then came Very Large Scale Integration (VLSI), with tens of thousands and hundreds of thousands of transistors. The trend continues today.

Although not everyone agrees that there is a fourth computer generation, those that do feel that it began in 1971, when IBM introduced its successors to the System/360 line of computers. These mainframe computers were called the System/370, and current-model IBM computers, although not called System/370s, evolved directly from these computers.

Minicomputers also proliferated during the fourth computer generation. The most popular lines were the DEC PDP-11 models and the DEC VAX, both of which are available in various models today.

Supercomputers first became prominent in the fourth generation. Although many companies, including IBM and CDC, developed high-speed computers for scientific work, it was not until Cray Research, Inc., introduced the Cray 1 in 1975 that supercomputers became significant. Today, supercomputers are an important computer classification.

Perhaps the most important trend that began in the fourth generation is the proliferation of microcomputers. As more and more transistors were put on silicon chips, it eventually became possible to put an entire computer processor, called a microprocessor, on a chip. The first computer to use microprocessors became available in the mid-1970s. The first microcomputer designed for personal use was the Altair, which was sold in 1975. The first Apple computer, marketed with the IBM PC in 1981. Today, microcomputers far outnumber all other types of computers combined.

Software development during the fourth computer generation started off with little change from the third generation. Operating systems were gradually improved, and new languages were designed. Database software became widely used during this time. The most important trend, however, resulted from the microcomputer revolution. Packaged software became widely available for microcomputers so that today most software is purchased, not developed from scratch.

1.2.5 Generationless Computers

We may have defined our last generation of computers and begun the era of generationless computers. Even though computer manufacturers talk of "fifth" and "sixth"-generation computers, this talk is more a marketing play than a reflection of reality.

Advocates of the concept of generationless computers say that even though technological innovations are coming in rapid succession, no single innovation is, or will be, significant enough to characterize another generation of computers.

New Words & Expressions

result in   导致，终于造成……结果
take delivery of   正式接过……
high-level language   高级语言
more and more   越来越多的
multiprogramming   n. 多道程序设计
virtual memory   虚拟内存
compatible   a. 兼容的；compatibility   n. 兼容性

air conditioning   空气调节
Navy lieutenant   海军上尉
mainframe   n. 主机，大型机
range from …to…   从……到……
time-share   n. 分时，时间共享
from scratch   从头开始
outnumber   vt. 数目超过，比……多

start off   v. 出发，开始   proliferate   v. 增生,扩散

Abbreviations

COBOL (Common Business-Oriented Language)   面向商业的通用语言
DEC (Digital Equipment Corporation)   美国数字设备公司
LSI (Large Scale Integrated Circuit)   大规模集成电路
VLSI (Very Large Scale Integrated Circuit)   超大规模集成电路

Notes

1. IBM introduced a line of mainframe computers called the System/360. IBM 公司推出了一个称为 System/360 的大型计算机系列，此处 line 指系列产品。

Reading Material: Classes of Computing Applications and Their Characteristics

Although a common set of hardware technologies is used in computers ranging from smart home appliances to cell phones to the largest supercomputers, these different applications have different design requirements and employ the core hardware technologies in different ways. Broadly speaking, computers are used in three different classes of applications.

Personal computers (PCs) are possibly the best known form of computing, which readers of this book have likely used extensively. Personal computers emphasize delivery of good performance to single users at low cost and usually execute third-party soft ware. This class of computing drove the evolution of many computing technologies, which is only about 35 years old!

Servers are the modern form of what were once much larger computers, and are usually accessed only via a network. Servers are oriented to carrying large workloads, which may consist of either single complex applications—usually a scientific or engineering application—or handling many small jobs, such as would occur in building a large web server. These applications are usually based on software from another source (such as a database or simulation system), but are often modified or customized for a particular function. Servers are built from the same basic technology as desktop computers, but provide for greater computing, storage, and input/output capacity. In general, servers also place a greater emphasis on dependability, since a crash is usually more costly than it would be on a single user PC.

Servers span the widest range in cost and capability. At the low end, a server may be little more than a desktop computer without a screen or keyboard and cost a thousand dollars. These low-end servers are typically used for file storage, small business applications, or simple web serving. At the other extreme are **supercomputers**, which at the present consist of tens of thousands of processors and many **terabytes** of memory, and cost tens to hundreds of millions of dollars. Supercomputers are usually used for high-end scientific and engineering calculations, such as weather forecasting, oil exploration, protein structure determination, and other large-scale problems. Although such supercomputers represent the peak of computing capability, they represent a relatively small fraction of the servers and a relatively small fraction of the overall computer market in terms of total revenue.

Embedded computers are the largest class of computers and span the widest range of applications and performance. Embedded computers include the microprocessors found in your car, the computers in a television set, and the networks of processors that control a modern airplane or cargo ship. Embedded computing systems are designed to run one application or one set of related applications that are normally integrated with the hardware and delivered as a single system; thus, despite the large number of embedded computers, most users never really see that they are using a computer!

Embedded applications often have unique application requirements that combine a minimum performance with stringent limitations on cost or power. For example, consider a music player: the processor need only be as fast as necessary to handle its limited function, and beyond that, minimizing cost and power are the most important objectives. Despite their low cost, embedded computers oft en have lower tolerance for failure, since the results can vary from upsetting (when your new television crashes) to devastating (such as might occur when the computer in a plane or cargo ship crashes). In consumer-oriented embedded applications, such as a digital home appliance, dependability is achieved primarily through simplicity—the emphasis is on doing one function as perfectly as possible. In large embedded systems, techniques of redundancy from the server world are oft en employed. Although this book focuses on general-purpose computers, most concepts apply directly, or with slight modifications, to embedded computers.

Many embedded processors are designed using *processor cores*, a version of a processor written in a hardware description language, such as Verilog or VHDL. The core allows a designer to integrate other application-specific hardware with the processor core for fabrication on a single chip.

New Words & Expressions

saerver   n. 服务器
workload   n. 工作量
low-end   a. 低端
stringent adj. 严格的
processor n. [计]处理器

fabrication   n. 兆兆(10^{12})字节；
database   n. 数据库
high-end   a. 高端
terabytes   n. 制造
Verilog   n. 一种硬件描述语言

Abbreviations

VHDL (VHSic hadware description language) 超高速集成电路硬件描述语言

科技英语的特点

比起非科技英语来，科技英语有四多，即复杂长句多、被动语态多、非谓语动词多、词性转换多。

一、复杂长句多

科技文章要求叙述准确，推理谨严，因此一句话里包含三四个甚至五六个分句的，并非少见。译成汉语时，必须按照汉语习惯破成适当数目的分句，才能条理清楚，避免洋腔洋调。这种复杂长句居科技英语难点之首，读者要学会运用语法分析方法来加以解剖，以便以短代长，化难为易。例如：

Factories will not buy machines unless they believe that the machine will produce goods that they are able to sell to consumers at a price that will cover all cost.

这是由一个主句和四个从句组成的复杂长句，只有进行必要的语法分析，才能正确理解和翻译。现试译如下：

除非相信那些机器造出的产品卖给消费者的价格足够支付所有成本，否则厂家是不会买那些机器的。

也可节译如下：

要不相信那些机器造出的产品售价够本，厂家是不会买的。

后一句只用了 24 个字，比前句 40 个字节约用字 40%，而对原句的基本内容无损。可见，只要吃透原文的结构和内涵，翻译时再在汉语上反复推敲提炼，复杂的英语长句，也是容易驾驭的。又如：

There is an increasing belief in the idea that the "problem solving attitude" of the engineer must be buttressed not only by technical knowledge and "scientific analysis" but that the engineer must also be aware of economics and psychology and, perhaps even more important, that he must understand the world around him.

这个长句由一个主句带三个并列定语从句构成，试译如下：

越来越令人信服的想法是：工程师不仅必须用技术知识和科学分析来加强解决问题的意向，而且也一定要了解经济学和心理学，而可能更为重要的是：必须懂得周围世界。

这两个例句初步说明了英语复杂长句的结构和译法。

二、被动语态多

英语使用被动语态大大多于汉语，如莎士比亚传世名剧《罗密欧与朱丽叶》中的一句就两次用了被动语态：

Juliet was torn between desire to keep Romeo near her and fear for his life, should his presence be detected.

朱丽叶精神上受到折磨，既渴望和罗密欧形影不离，又担心罗密欧万一让人发现，难免有性命之忧。

科技英语更是如此，有三分之一以上用被动语态。例如：

(a) No work can be done without energy.

译文：没有能量决不能做功。

(b) All business decisions must now be made in the light of the market.

译文：所有企业现在必须根据市场来作出决策。

(c) Automobiles may be manufactured with computer-driven robots or put together almost totally by hand.

译文：汽车可以由计算机操纵的机器人来制造，或者几乎全部用手工装配。

以上三例都用被动语态。但译成汉语时都没有用被动语态，以便合乎汉语传统规范。例(c)

的并列后句，其谓语本应是 may be put together。put 是三种变化形式一样的不规则动词，在这里是过去分词，由于修辞学上避免用词重复出现的要求，略去了 may be 两词，所以并非现在时，而是被动语态。

科技英语之所以多用被动语态，为的是要强调所论述的客观事物（四例中的 work, necessaries, business decisions, automobiles），因此放在句首，作为句子的主语，以突出其重要性。

三、非谓语动词多

英语每个简单句中，只能用一个谓语动词，如果读到几个动作，就必须选出主要动作当谓语，而将其余动作用非谓语动词形式，才能符合英语语法要求。

非谓语动词有三种：动名词、分词（包括现在分词和过去分词）和不定式。例如：

(a) 要成为一个名符其实的内行，需要学到老。

这句中，有"成为""需要"和"学"三个表示动作的词，译成英语后为：

To be a true professional requires lifelong learning.

可以看出，选好"需要"（require）作为谓语，其余两个动作："成为"用不定式形式 to be，而"学"用动名词形式 learning，这样才能符合英语语法要求。

(b) 任何具有重量并占有空间的东西都是物质。

这句包含"是"（在英语中属于存在动词）、"具有"和"占有"三个动作，译成英语为：

Matter is anything having weight and occupying space.

将"是"（is）当谓语（系动词），而"具有"（having）和"占有"（occupying）处理为现在分词，连同它们的宾语 weight 和 space 分别构成现在分词短语作为修饰名词 anything 的定语。

(c) 这门学科为人所知的两大分支是无机化学和有机化学。

这句有"为人所知"和"是"两个动词，译成英语后为：

The two great divisions of this science known are inorganic chemistry and organic chemistry.

这里将"是"（are）作为谓语系动词，而将"为人所知"（known）处理为过去分词。

上述三例分别列举了三种非谓语动词的使用情况。其必要性都是为了英语语法上这条铁定的要求：每个简单句只允许有一个谓语动词。这就是英语为什么不同于其他语言，有非谓语动词，而且用得十分频繁的原因。

四、词性转换多

英语单词有不少是多性词，即既是名词，又可用作动词、形容词、介词或副词，字形无殊，功能各异，含义也各不相同，如不仔细观察，必致谬误。例如：

(a) above

介词：above all (things) 首先，最重要的是

形容词：for the above reason 由于上述理由

副词：As (has been) indicated above 如上所指出

(b) light

名词：（启发）in (the) light of 由于，根据；

（光）high light(s) 强光，精华；

（灯）safety light 安全指示灯

形容词：（轻）light industry 轻工业；

（明亮）light room 明亮的房间；

（淡）light blue 淡蓝色；

（薄）light coating 薄涂层

动词：（点燃）light up the lamp 点灯

副词：（轻快）travel light 轻装旅行

（容易）light come, light go 来得容易去得快

诸如此类的词性转换，在德、俄等西方语言中是少有的，而科技英语中却屡见不鲜，几乎每个技术名词都可转换为同义的形容词。词性转换增加了英语的灵活性和表现力，读者必须从上下文判明用词在句中是何种词性，含义如何，才能对全句得到正确无误的理解。

我们在科技翻译实践中，要充分体现以上各个特点，重视信息传递，注意调整句式、篇章，以使译文叙述条理、逻辑连贯，同时还要注意准确使用科技术语。

Exercises

I. Answer the following questions

1. When was the modern computer invented?
2. What are major characteristics of the four generations of modern computers?
3. Describe the near-future supercomputer directions.
4. What are basic characteristics of modern computers?

II. Write a summary of section 1.2 about computer generations in 300 words.

III. Talk about the trends of computer hardware and software.

Chapter 2　Basic Organization of Computers

学习指导

计算机主要由中央处理器、存储设备以及输入、输出设备等组成。通过本章学习，读者应掌握以下内容：
- 掌握计算机结构与硬件的主要术语；
- 掌握计算机的组成与各部分的功能，并能用英语表述；
- 掌握专业词汇的构成规律，特别是常用词缀及复合词的构成。

2.1　Introduction

In this chapter, we examine the organization of basic computer systems. A simple computer has three primary subsystems. The central processing unit, or CPU, performs many operations and controls the computer. A microprocessor usually serves as the computer's CPU. The memory subsystem is used to store programs being executed by the CPU, along with the program's data. The input/output, or I/O, subsystem allows the CPU to interact with input and output devices, such as the keyboard and monitor of a personal computer, or the keypad and digital display of a microwave oven.

Most computer systems, from the embedded controllers found in automobiles and consumer appliances to personal computers and mainframes, have the same basic organization. This organization has three main components: the CPU, the memory subsystem, and the I/O subsystem. The generic organization of these components is shown in Figure 2-1.

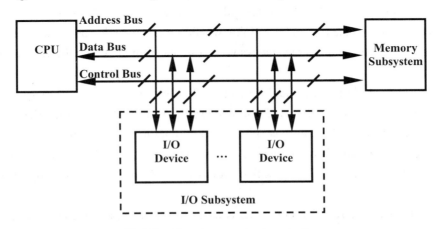

Fig.2-1　Generic computer organization

In this chapter, we first describe the system buses used to connect the components in the computer system. Then we examine the instruction cycle, the sequence of operations that occurs within the computer as it fetches, decodes, and executes an instruction.

New Words & Expressions

subsystem　n. 子系统

microprocessor　n. [计]微处理器

sequence　n. 时序，序列

decode　vt. 解码，译解

operation　n. 操作，运算，执行命令(计)

system buses　系统总线

fetch　vt. 取数，取指令

instruction　n. 指令

Abbreviations

CPU(Central Processing Unit)　中央处理器

I/O(Input/Output)　输入输出(设备)

2.2　System Buses

Physically, a bus is a set of wires. The components of the computer are connected to the buses. To send information from one component to another, the source component outputs data onto the bus. The destination component then inputs this data from the bus. As the complexity of a computer system increases, it becomes more efficient (in terms of minimizing connections) at using buses rather than direct connections between every pair of devices. Buses use less space on a circuit board and require less power than a large number of direct connections. They also require fewer pins on the chip or chips that comprise the CPU.

The system shown in Figure 2-1 has three buses. The uppermost bus in this figure is the address bus. When the CPU reads data or instructions from or writes data to memory, it must specify the address of the memory location it wishes to access. It outputs this address to the address bus; memory inputs this address from the address bus and use it to access the proper memory location. Each I/O devices, such as a keyboard, monitor, or disk drive, has a unique address as well. When accessing an I/O device, the CPU places the address of the device on the address bus. Each device can read the address off of the bus and determine whether it is the device being accessed by the CPU. Unlike the other buses, the address bus always receives data from the CPU; the CPU never reads the address bus.

Data is transferred via the data bus. When the CPU fetches data from memory, it first outputs the memory address on its address bus. Then memory outputs the data onto the data bus; the CPU can then read the data from the data bus. When writing data to memory, the CPU first outputs the address onto the address bus, then outputs the data onto the data bus. Memory then reads and stores the data at the proper location. The processes for reading data from and writing data to the I/O devices are similar.

The control bus is different from the other two buses. The address bus consists of n lines, which combine to transmit one n-bit address value. Similarly, the lines of the data bus work together to transmit a single multibit value. In contrast, the control bus is a collection of individual control signals. These signals indicate whether data is to be read into or written out of the CPU, whether the CPU is accessing memory or an I/O device, and whether the I/O device or memory is ready to transfer data. Although this bus is shown as bidirectional in Figure 2-1, it is really a collection of (mostly) unidirectional signals. Most of these signals are output from the CPU to the memory and I/O subsystems, although a few are output by these subsystems to the CPU. We examine these signals in more detail when we look at the instruction cycle and the subsystem interface.

A system may have a hierarchy of buses. For example, it may use its address, data, and control buses to access memory, and an I/O controller. The I/O controller, in turn, may access all I/O devices using a second bus, often called an I/O bus or a local bus.

New Words & Expressions

pins　n. 插脚，管脚
uppermost　adj. 最高的；adv. 在最上
data bus　数据总线
multibit　多位
unidirectional　单向的
I/O bus　输入输出总线

address bus　地址总线
control bus　控制总线
via　prep. 经，通过，经由
bidirectional　双向的
hierarchy　n. 层次，层级
local bus　n. 局域总线

2.3　Instruction Cycle

The instruction cycle is the procedure a microprocessor goes through to process an instruction. First the microprocessor fetches, or reads, the instruction from memory. Then it decodes the instruction, determining which instruction it has fetched. Finally, it performs the operations necessary to execute the instruction. (Some people also include an additional element in the instruction cycle to store results. Here, we include that operation as part of the execute function.) Each of these functions—fetch, decode, and execute—consists of a sequence of one or more operations.

Let's start where the computer starts, with the microprocessor fetching the instruction from memory. First, the microprocessor places the address of the instruction on to the address bus. The memory subsystem inputs this address and decodes it to access the sired memory location. (We look at how this decoding occurs when we examine the memory subsystem in more detail later in this chapter.)

After the microprocessor allows sufficient time for memory to decode the address and access the requested memory location, the microprocessor asserts a READ control signal. The READ signal is a signal on the control bus which the microprocessor asserts when it is ready to read data from memory or an I/O device. (Some processors have a different name for this signal, but all microprocessors have a signal to perform this function.) Depending on the microprocessor, the READ signal may be active high (asserted - 1) or active low (asserted - 0).

When the READ signal is asserted, the memory subsystem places the instruction code to be fetched onto the computer system's data bus, The microprocessor then inputs this data from the bus and stores it in one of its internal registers. At this point, the microprocessor has fetched the instruction.

Next, the microprocessor decodes the instruction. Each instruction may require a different sequence of operations to execute the instruction. When the microprocessor decodes the instruction, it determines which instruction it is in order to select the correct sequence of operations to perform. This is done entirely within the microprocessor; it does not use the system buses.

Finally, the microprocessor executes the instruction. The sequence of operations to execute the instruction varies from instruction to instruction. The execute routine may read data from memory, write data to memory, read data from or write data to an I/O device, perform only operations within the CPU, or perform some combination of these operations. We now look at how the computer performs these operations from a system perspective.

To read data from memory, the microprocessor performs the same sequence of operations it uses to fetch an instruction from memory. After all, fetching an instruction is simply reading it from memory. Figure 2-2(a) shows the timing of the operations to read data from memory.

In Figure 2-2, notice the top symbol, CLK. This is the computer system clock; the microprocessor uses the system clock to synchronize its operations. The microprocessor places the address onto the bus at the beginning of a clock cycle, a 0/1 sequence of the system clock. One clock cycle later, to allow time for memory to decode the address and access its data, the microprocessor asserts the READ Signal. This causes memory to place its data onto the system data bus. During this clock cycle, the microprocessor reads the data off the system bus and stores it in one of its registers. At the end of the clock cycle it removes the address from the address bus and deasserts the READ signal. Memory then removes the data from the data bus, completing the memory read operation.

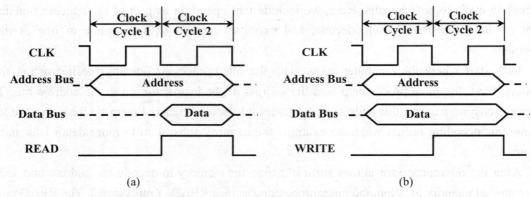

Fig.2-2　Timing diagram for (a) memory read and (b) memory write

The timing of the memory write operation is shown in Figure 2-2(b). The processor places the address and data onto the system buses during the first clock cycle. The microprocessor then asserts a WRITE control signal (or its equivalent) at the start of the second clock cycle. Just as the READ signal causes memory to read data, the WRITE signal triggers memory to store data. Some time during this cycle, memory writes the data on the data bus to the memory location whose address is on the address bus. At the end of this cycle, the processor completes the memory write operation by removing the address and data from the system buses and deasserting the WRITE signal.

The I/O read and write operations are similar to the memory read and write operations. A processor may use either memory mapped I/O or isolated I/O. If the processor supports memory mapped I/O, it follows the same sequences of operations to input or output data as to read data from or write data to memory, the sequences shown in Figure 2-2. (Remember, in memory mapped I/O, the processor treats an I/O port as a memory location, so it is reasonable to treat an I/O data access the same as a memory access.) Processors that use isolated I/O follow the same process but have a second control signal to distinguish between I/O and memory accesses. (CPUs that use isolated I/O can have a memory location and an I/O port with the same address, which makes this extra signal necessary.)

Finally, consider instructions that are executed entirely within the microprocessor. The INAC instruction of the Relatively Simple CPU, and the MOV r1, r2 instruction of the 8085 microprocessor, can be executed without accessing memory or I/O devices. As with instruction decoding, the execution of these instructions does not make use of the system buses.

New Words & Expressions

instruction cycle 指令周期
register n. 寄存器
timing n. 定时；时序；时间选择
assert vt. 主张，发出
trigger vt. 引发，引起，触发

memory map n. [计]内存
port n. 端口
synchronize vt. 使…同步
deassert vt. 撤销
map v. 映射

2.4 CPU ORGANIZATION

The CPU controls the computer. It fetches instructions from memory, supplying the address and control signals needed by memory to access its data. The CPU decodes the instruction and controls the execution procedure. It performs some operations internally, and supplies the address, data, and control signals needed by memory and I/O devices to execute the instruction. Nothing happens in the computer unless the CPU causes it to happen.

Internally, the CPU has three sections, as shown in Figure 2-3. The register sections, as its name implies, includes a set of registers and a bus or other communication mechanism. The registers in a processor's instruction set architecture are found in this section of the CPU. The system address and data buses interact with this section of the CPU. The register section also contains other registers that are not directly accessible by the programmer. The relatively simple CPU includes registers to latch the address being accessed in memory and a temporary storage register, as well as other registers that are not a part of its instruction set architecture.

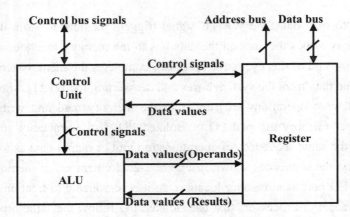

Fig.2-3　CPU Internal Organization

During the fetch portion of the instruction cycle, the processor first outputs the address of the instruction onto the address bus. The processor has a register called the program counter; the CPU keeps the address of the next instruction to be fetched in this register. Before the CPU outputs the address onto the system's address bus, it retrieves the address from the program counter register. At the end of the instruction fetch, the CPU reads the instruction code from the system data bus. It stores this value in an internal register, usually called the instruction register or something similar.

The arithmetic logic unit or ALU performs most arithmetic and logical operations, such as adding or ANDing values. It receives its operands from the register section of the CPU and stores its results back in the register section. Since the ALU must complete its operations within a single clock cycle, it is constructed using only combinatorial logic. The ADD instructions in the relatively simple CPU and the 8085 microprocessor use the ALU during their executions.

Just as the CPU controls the computer (in addition to its other functions), the control unit controls the CPU. This unit generates the internal control signals that cause registers to load data, increment or clear their contents, and output their contents, as well as cause the ALU to perform the correct function. These signals are shown as control signals in Figure 2-3. The control unit receives some data values from the register unit, which it uses to generate the control signals. This data includes the instruction code and the values of some flag registers. The control unit also generates the signals for the system control bus, such as the READ, WRITE, and $IO/\overline{M}$ signals. A microprocessor typically performs a sequence of operations to fetch, decode, and execute an instruction. By asserting these internal and external control signals in the proper sequence, the control unit causes the CPU and the rest of the computer to perform the operations needed to correctly process instructions.

This description of the CPU is incomplete. Current processors have more complex features that improve their performance. One such mechanism, the instruction pipeline, allows the CPU to fetch one instruction while simultaneously executing another instruction.

New Words & Expressions

latch   v. 闭锁，锁存
instruction register   指令寄存器
increment   n. 增量，加 1
pipeline   n. 流水线

program counter   程序计数器
operand   n. 操作数
flag register   标志寄存器
microsequenced   微层序的

Abbreviations

ALU (Arithmetic Logic Unit)   算术逻辑单元

Reading Material: Eight Great Ideas in Computer Architecture

We now introduce eight great ideas that computer architects have been invented in the last 60 years of computer design. These ideas are so powerful they have lasted long after the first computer that used them, with newer architects demonstrating their admiration by imitating their predecessors. These great ideas are themes that we will weave through this and subsequent chapters as examples arise. To pointout their influence, in this section we introduce icons and highlighted terms that represent the great ideas and we use them to identify the nearly 100 sections of the book that feature use of the great ideas.

Design for Moore's Law

The one constant for computer designers is rapid change, which is driven largely by **Moore's Law**. It states that integrated circuit resources double every 18－24 months. Moore's Law resulted from a 1965 prediction of such growth in IC capacity made by Gordon Moore, one of the founders of Intel. As computer designs can take years, the resources available per chip can easily double or quadruple between the start and finish of the project. Like a skeet shooter, computer architects must anticipate where the technology will be when the design finishes rather than design for where it starts. We use an "up and to the right" Moore's Law graph to represent designingfor rapid change.

Use Abstraction to Simplify Design

Both computer architects and programmers had to invent techniques to make themselves more productive, for otherwise design time would lengthen as dramatically as resources grew by Moore's Law. A major productivity technique for hardware and soft ware is to use **abstractions** to represent the design at different levels of representation; lower-level details are hidden to offer a simpler model at higher levels. We'll use the abstract painting icon to represent this second great idea.

Make the Common Case Fast

Making the **common case fast** will tend to enhance performance better than optimizing the rare case. Ironically, the common case is often simpler than the rare case and hence is often easier to enhance. This common sense advice implies that you know what the common case is, which is only possible with careful experimentation and measurement. We use a sports car as the icon for making the common case fast, as the most common trip has one or two passengers, and it's surely easier to make a fast sports car than a fast minivan!

Performance via Parallelism

Since the dawn of computing, computer architects have offered designs that get more performance by performing operations in parallel. We'll see many examples of parallelism in this book. We use multiple jet engines of a plane as our icon for **parallel performance**.

Performance via Pipelining

A particular pattern of parallelism is so prevalent in computer architecture that it merits its own name: **pipelining**. For example, before fire engines, a "bucket brigade" would respond to a fire, which many cowboy movies show in response to a dastardly act by the villain. The townsfolk form a human chain to carry a water source to fire, as they could much more quickly move buckets up the chain instead of individuals running back and forth. Our pipeline icon is a sequence of pipes, with each section representing one stage of the pipeline.

Performance via Prediction

Following the saying that it can be better to ask for forgiveness than to ask for permission, the final great idea is prediction. In some cases it can be faster on average to guess and start working rather than wait until you know for sure, assuming that the mechanism to recover from a misprediction is not too expensive and your prediction is relatively accurate. We use the fortune-teller's crystal ball as our prediction icon.

Hierarchy of Memories

Programmers want memory to be fast, large, and cheap, as memory speed often shapes performance, capacity limits the size of problems that can be solved, and the cost of memory today is oft en the majority of computer cost. Architects have found that they can address these conflicting demands with a hierarchy of memories, with the fastest, smallest, and most expensive memory per bit at the top of the hierarchy and the slowest, largest, and cheapest per bit at the bottom. Caches give the programmer the illusion that main memory is nearly as fast as the top of the hierarchy and nearly as big and cheap as the bottom of the hierarchy. We use a layered triangle icon to represent the memory hierarchy. The shape indicates speed, cost, and size: the closer to the top, the faster and more expensive per bit the memory; the wider the base of the layer, the bigger the memory.

Dependability via Redundancy

Computers not only need to be fast; they need to be dependable. Since any physical device can fail, we make systems dependable by including redundant components that can take over when a failure occurs and to help detect failures. We use the tractor-trailer as our icon, since the dual tires on each side of its rear axels allow the truck to continue driving even when one tire fails. (Presumably, the truck driver heads immediately to a repair facility so the flat tire can be fixed, thereby restoring redundancy!)

New Words & Expressions

admiration n. 钦佩；赞赏；羡慕
skeet n. 双向

theme n. 主题
shooter n. 枪；射手

abstractions　n. 抽象，抽象化
pipeline　v. 管道输送
dastardly　ad. 懦弱的，卑鄙的
hierarchy　n. 层次体系
redundancy　n. 过剩，冗余
axels　n. 轮轴

parallelism　n. 平行，并行
bucket brigade　n. 救火队列
villain　n. 恶棍；歹徒；坏人，罪犯
cache　n. 快速缓冲贮存区
rear　n. 后部，背

计算机英语专业词汇的构成

英语的词汇构成有很多种，真正英语的基本词汇是不多的，很大部分词汇属于构成型词汇。这里，仅介绍在专业英语中遇到的专业词汇及其构成。目前，各行各业都有一些自己领域的专业词汇，有的是随着本专业发展应运而生的，有的是借用公共英语中的词汇，有的是借用外来语言词汇，有的则是人为构造成的词汇。

一、派生词（derivation）

这类词汇非常多，它是根据已有的词加上某种前后缀，或以词根生成、或以构词成分形成新的词。科技英语词汇中有很大一部分来源于拉丁语、希腊语等外来语，有的是直接借用，有的是在它们之上不断创造出新的词汇。这些词汇的构词成分（前缀、后缀、词根等）较固定，构成新词以后便于读者揣度词义，易于记忆。

1.1 前缀

采用前缀构成的单词在计算机专业英语中占了很大比例，通过下面的实例可以了解这些常用的前缀构成的单词。

multi- 多
multiprogram 多道程序
multimedia 多媒体
multiprocessor 多处理器
multiplex 多路复用
multiprotocol 多协议
inter- 相互、在...间
interface 接口、界面
interlace 隔行扫描
interlock 联锁
internet 互联网络（因特网）
interconnection 互联

hyper- 超级
hypercube 超立方
hypercard 超级卡片
hypermedia 超媒体
hypertext 超文本
hyperswitch 超级交换机
micro- 微型
microprocessor 微处理器
microkernel 微内核
microcode 微代码
microkid 微机迷
microchannel 微通道

super 超级
superhighway 超级公路
superpipeline 超流水线
superscalar 超标量
superset 超集
superclass 超类
tele- 远程的
telephone 电话
teletext 图文电视
telemarketing 电话购物
telecommuting 家庭办公
teleconference 远程会议

单词前缀还有很多，其构成可以同义而不同源（如拉丁、希腊），可以互换，例如：

multi, poly 相当于 many　　如：multimedia 多媒体，polytechnic 各种工艺的
uni, mono 相当于 single　　如：unicode 统一的字符编码标准，monochrome 单色
bi, di 相当于 twice　　如：bichloride 双氯的，dichloride 二氯化物
equi,iso 相当于 equal　　如：equality 等同性, isoline 等值线

simili, homo 相当于 same   如：similarity 类似，homogeneous 同类的
semi, hemi 相当于 half   如：semiconductor 半导体，hemicycle 半圆形
hyper, super 相当于 over   如：hypertext 超文本，superscalar 超标量体系结构

1.2 后缀

后缀是在单词后部加上构词结构，形成新的单词。如：

-scope 探测仪器	-meter 计量仪器	-graph 记录仪器
baroscope 验压器	barometer 气压表	barograph 气压记录仪
telescope 望远镜	telemeter 测距仪	telegraph 电报
spectroscope 分光镜	spectrometer 分光仪	spectrograph 分光摄像仪
-able 可能的	-ware 件（部件）	-ity 性质
enable 允许、使能	hardware 硬件	reliability 可靠性
disable 禁止、不能	software 软件	availability 可用性
programmable 可编程的	firmware 固件	accountability 可核查性
portable 便携的	groupware 组件	integrity 完整性
scalable 可缩放的	freeware 赠件	confidentiality 保密性

二、复合词（compounding）

复合词是科技英语中另一大类词汇，其组成面广，通常分为复合名词、复合形容词、复合动词等。复合词通常以小横杠"-"连接单词构成，或者采用短语构成。有的复合词进一步发展，去掉了小横杠，并经过缩略成为另一类词类，即混成词。复合词的实例有：

-based 基于，以……为基础	-centric 以……为中心的
rate-based 基于速率的	client-centric 以客户为中心的
credit-based 基于信誉的	user-centric 以用户为中心的
file-based 基于文件的	host-centered 以主机为中心的
Windows-based 以 Windows 为基础的	
-oriented 面向……的	-free 自由的，无关的
object-oriented 面向对象的	lead-free 无线的
market-oriented 市场导向	jumper-free 无跳线的
process-oriented 面向进程的	paper-free 无纸的
thread-oriented 面向线程的	charge-free 免费的
info- 信息，与信息有关的	info-channel 信息通道
info-tree 信息、树	info-world 信息世界
info-sec 信息安全	

其他

point-to-point 点到点	point-and-click 点击
plug-and-play 即插即用	drag-and-drop 拖放
easy-to-use 易用的	line-by-line 逐行
off-the-shelf 现成的	store-and-forward 存储转发
peer-to-peer 对等的	operator-controllable 操作员可控制的

leading-edge 领先的　　　　　　over-hyped 过度宣扬的
end-user 最终用户　　　　　　front-user 前端用户
sign-on 登录　　　　　　　　sign-of 取消
pull-down 下拉　　　　　　　pull-up 上拉
pop-up 弹出

此外，以名词 + 动词-ing 构成的复合形容词形成了一种典型的替换关系，即可以根据需要在结构中代入同一词类而构成新词，它们多为动宾关系。如：

man-carrying aircraft 载人飞船　　　earth-moving machine 推土机
time-consuming operation 耗时操作　ocean-going freighter 远洋货舱

然而，必须注意，复合词并非随意可以构造，否则会形成一种非正常的英语句子结构。虽然上述例子给出了多个连接单词组成的复合词，但不提倡这种冗长的复合方式。对于多个单词的非连线形式，要注意其顺序和主要针对对象。此外还应当注意，有时加连字符的复合词与不加连字符的词汇词意是不同的，必须通过文章的上下文推断。如：

force-feed 强迫接受（vt.），而 force feed 则为"加压润滑"。

随着词汇的专用化，复合词中间的连接符被省略掉，形成了一个单词，例如：

videotape 录像带　　　fanin 扇入　　　　fanout 扇出
online 在线　　　　　onboard 在板　　　login 登录
logout 撤消　　　　　pushup 拉高　　　popup 弹出

三、混成词（blending）

混成词不论在公共英语还是科技英语中都大量出现，也有人将它们称为缩合词（与缩略词区别）、融会词，它们多是名词，也有地方将其作为动词用，对这类词汇可以通过其构词规律和词素进行理解。这类词汇将两个单词的前部拼接、前后拼接或者将一个单词前部与另一词拼接构成新的词汇，实例有：

brunch (breakfast + lunch) 早中饭　　　　smog (smoke +fog) 烟雾
codec (coder+decoder) 编码译码器　　　　compuser (computer+user) 计算机用户
transeiver (transmitter+receiver) 收发机　　syscall (system+call) 系统调用
mechatronics (mechanical+electronic) 机械电子学
calputer (calculator+computer) 计算器式电脑

四、缩略词（shortening）

缩略词是将较长的英语单词取其首部或者主干构成与原词同义的短单词，或者将组成词汇短语的各个单词的首字母拼接为一个大写字母的字符串。随着科技发展，缩略词在文章索引、前序、摘要、文摘、电报、说明书、商标等科技文章中频繁采用。对计算机专业来说，在程序语句、程序注释、软件文档、文件描述中也采用了大量的缩略词作为标识符、名称等等。缩略词的出现方便了印刷、书写、速记、以及口语交流等，但也同时增加了阅读和理解的困难。

缩略词开始出现时，通常采用破折号、引号或者括号将它们的原形单词和组合词一并列出，久而久之，人们对缩略词逐渐接受和认可，作为注释性的后者也就消失了。在通常情况下，缩略词多取自各个组合字（虚词除外）的首部第一、二字母。缩略词也可能有形同而义异的情况。如果遇到这种情况，翻译时应当根据上下文确定词意，并在括号内给出其原形组合词汇。缩略词可以分为如下几种。

4.1 压缩和省略

将某些太长、难拼、难记、使用频繁的单词压缩成一个短小的单词，或取其头部、或取其关键音节。如：

flu=influenza 流感　　　　lab=laboratory 实验室　　　math=mathematics 数学
iff=if only if 当且仅当　　　rhino=rhinoceros 犀牛　　　ad=advertisement 广告

4.2 缩写（acronym）

将某些词组和单词集合中每个实意单词的第一或者首部几个字母重新组合，组成为一个新的词汇，作为专用词汇使用。在应用中它形成三种类型，即：

（1）通常以小写字母出现，并作为常规单词

radar (radio detecting and ranging) 雷达
laser (light amplification by stimulated emission of radiation) 激光
sonar (sound navigation and ranging) 声纳
spool (simultaneous peripheral operation on line) 假脱机

（2）以大写字母出现，具有主体发音音节

BASIC (Beginner's All-purpose Symbolic Instruction Code) 初学者通用符号指令代码
FORTRAN (Formula Translation) 公式翻译
COBOL (Common Business Oriented Language) 面向商务的通用语言

（3）以大写字母出现，没有读音音节，仅为字母头缩写

ADE (Application Development Environment) 应用开发环境
PCB (Process Control Block) 进程控制块
CGA (Color Graphics Adapter) 彩色图形适配器
DBMS (Data Base Management System) 数据库管理系统
FDD (Floppy Disk Device) 软盘驱动器
MBPS (Mega Byte Per Second) 每秒兆字节
Mbps (Mega Bits Per Second) 每秒兆字位
RISC (Reduced Instruction Set Computer) 精简指令集计算机
CISC (Complex Instruction Set Computer) 复杂指令集计算机

五、借用词

借用词一般来自厂商名、商标名、产品代号名、发明者名、地名等，它通过将普通公共英语词汇演变成专业词意而实现。有的则是将原来已经有的词汇赋予新的含义。例如：

woofer 低音喇叭　　　　tweeter 高音喇叭　　　　flag 标志、状态
cache 高速缓存　　　　semaphore 信号量　　　　firewall 防火墙
mailbomb 邮件炸弹　　　scratch pad 高速缓存　　　fitfall 专用程序入口

在现代科技英语中借用了大量的公共英语词汇、日常生活中的常用词汇，而且，以西方特有的幽默和结构讲述科技内容。这时，读者必须在努力扩大自己专业词汇的同时，也要掌握和丰富自己的生活词汇，并在阅读和翻译时正确采用适当的含义。

Exercises

I. Answer the following questions

1. Describe the organization of basic computer systems.
2. How does a processor process an instruction?
3. How many sections are there in a CPU, and what are their functions?

II. The eight great ideas in computer architecture are similar to ideasfrom other fields. Match the eight ideas from computer architecture, "Design forMoore's Law" "Use Abstraction to Simplify Design" "Make the Common CaseFast" "Performance via Parallelism" "Performance via Pipelining" "Performancevia Prediction" "Hierarchy of Memories", and "Dependability via Redundancy" to the following ideas from other fields:

    a. Assembly lines in automobile manufacturing
    b. Suspension bridge cables
    c. Aircraft and marine navigation systems that incorporate wind information
    d. Express elevators in buildings
    e. Library reserve desk
    f. Increasing the gate area on a CMOS transistor to decrease its switching time
    g. Adding electromagnetic aircraft catapults (which are electrically-poweredas opposed to current steam-powered models), allowed by the increased powergeneration offered by the new reactor technology
    h. Building self-driving cars whose control systems partially rely on existing sensorsystems already installed into the base vehicle, such as lane departure systems andsmart cruise control systems

Chapter 3 Binary System and Boolean Algebra

学习指导

在日常生活中,我们普遍采用的十进制系统,而在计算机领域广泛采用的是二进制系统。布尔代数法则在集合运算、二进制运算、逻辑运算的普遍应用,确立了布尔代数在电子计算机发展中的中心地位。通过本章的学习,读者应该掌握以下内容:
- 十进制和二进制系统的特点及其主要术语;
- 布尔代数与集合运算的关系,布尔代数中的主要定理的表述;
- 掌握科技英语中数学公式的读法。

The decimal system for counting has been so widely adopted throughout our present civilization that we rarely consider the possibilities of other number systems. Nevertheless, it is not reasonable to expect a system based on the number of fingers we possess to be the most efficient number system for machine construction. The fact is that a little-used but very simple system, the binary number system, has proved the most natural and efficient system for machine use.

3.1 The Decimal System

Our present system of numbers has 10 separate symbols 0, 1, 2, 3,···, 9, wich are called Arabic numerals. We would be forced to stop at 9 or to invent more symbols if it were not for the use of positional notation. An example of earlier types of notation can be found in Roman numeral, which are essentially additive: III=I+I+I, XXV=X+X+V. New symbols (X, C, M, etc.) were used as the numbers increased in value. Thus V rather than IIIII=5. The only importance of position in Roman numerals lies in whether a symbol precedes or follows another symbol (IV=4 and VI=6). The clumsiness of this system can easily be seen if we try to multiply XII by XIV. Calculating with Roman numerals was so difficult that early mathematicians were forced to perform arithmetic operations almost entirely on abaci, or counting boards, translating their results back into Roman-number form. Pencil-and-paper computations are unbelievably intricate and difficult in such systems. In fact, the ability to perform such operations as addition and multiplication was considered a great accomplishment in earlier civilization.

The great beauty and simplicity of our number system can now be seen. It is necessary to learn only the basic numerals and the positional notation system in order to count to any desired figure. After memorizing the addition and multiplication tables and learning a few simple rules, it is possible to perform all arithmetic operations. Notice the simplicity of multiplying 12×14 using the present system:

```
        14
    ×   12
        28
       14
       168
```

The actual meaning of the number 168 can be seen more clearly if we notice that it is spoken as "one hundred and sixty-eight". Basically, the number is a contraction of (1×100)+(6×10)+8. The important point is that the value of each digit is determined by its position. For example, the 2 in 2,000 has a different value than the 2 in 20. We show this verbally by saying "two thousand" and "twenty". Different verbal representations have been invented for numbers from 10 to 20 (eleven, twelve), but from 20 upward we break only at powers of 10 (hundreds, thousands, millions, billions). Written numbers are always contracted, however, and only the basic 10 numerals are used regardless of the size of the integer written. The general rule for representing numbers in the decimal system using positional notation is as follows:

$$a_{n-1}10^{n-1} + a_{n-2}10^{n-2} + \cdots + a_0 10^0$$

The integer digit in different position is expressed as a_{n-1}, a_{n-2}, $\cdots$, a_0 where "n" is the number of digits to the left of the decimal point.

The base, or radix of a number system is defined as the number of different digits which can occur in each position in the number system. The decimal number system has a base, or radix, of 10. This means that the system has 10 different digits (0, 1, 2, $\cdots$, 9), any one of which may be used in each position in a number. History records the use of several other number systems. The quinary system, which has 5 for its base, was prevalent among Eskimos and North American Indians. Examples of the duodecimal system (base 12) may be seen in clocks, inches and feet and in dozens or grosses.

New Words & Expressions:

decimal system　十进制
verbal representation　口头表达式
quinary system　五进制
stop at　停在
in fact　事实上
Roman-number　n. 罗马数字
regardless of　a. 不管，不顾
gross　n. 罗，为一计数单位，1 罗=12 打
abacus　n. 算盘
radix　n. 根，基数
duodecimal system　十二进制
lie in　在于
in order to　为了
pencil-and-paper　纸和笔的
contraction　n. 收缩，缩写式，紧缩
positional notation　n. 位置记数法

3.2　The Binary System

A seventeenth-century German mathematician Gottfried Wilhelm von Leibniz, was an advocate of the binary number system which has 2 for a base, using only the symbols 0 and 1. It seems strange for an eminent mathematician to advocate such a simple number system; it should be noted that he was also a philosopher. Leibniz's reasons for advocating the binary system seem to have been mystical. He felt there was great beauty in the analogy between zero, representing the void, and one, representing the Deity.

Regardless of how good Leibniz's reasons for advocating it were, the binary system has become very popular in the last decade. Present-day digital computers are constructed to operate in binary or binary-coded number system, and present indications are that future machines will also be constructed to operate in these system.

The basic elements in early computers were relays and switches. The operation of a switch, or relay, can be seen to be essentially binary in nature. That is the switch is either on (1) or off (0). The principal circuit elements in more modern computers are transistors similar to those used in radios and television sets. The desire for reliability led designers to use these devices so that they were essentially in one of two states, fully conduction or non-conducting. A simple analogy may be made between this type of circuit and an electric light. At any given time the light (or transistor) is either on (conducting) or off (not conduction). Even after a bulb is old and weak, it is generally easy to tell if it is on or off. The same sort of thing may be seen in radios. As a radio ages, the volume generally decreases, and we compensate by turning the volume control up. Even when the radio becomes very weak, however, it is still possible to tell easily whether it is on or off.

Because of the large number of electronic parts used in computers, it is highly desirable to utilize them in such a manner that slight changes in their characteristics will not affect their performance. The best way of accomplishing this is using circuits which are basically bistable (having two possible states).

New Words & Expressions:

Binary system　二进制　　　　　　　　Binary coded system　二进制编码系统
bistable　a. 双稳态　　　　　　　　　　relay　n. 继电器
regardless of　无论，不管　　　　　　　lead to　导致，通向
in nature　本质上，实际上，事实上　　　similar to　与……同类，与……相似

3.3　Boolean Algebra

The concept of a Boolean algebra was first proposed by the English mathematician George Boole in 1847. Since that time, Boole's original conception has been extensively developed and refined by algebraists and logicians. The relationships among Boolean algebra, set algebra, logic, and binary arithmetic have given Boolean algebras a central role in the development of electronic digital computers.

The most intuitive development of Boolean algebras arises from the concept of a set algebra. Let S={a, b, c} and T={a, b, c, d, e} be two sets consisting of three and five elements, respectively. We say that S is a subset of T, since every element of S (namely, a, b, and c) belongs to T. Since T has five elements, there are 2^5 subsets of T, for we may choose any individual element to be included or omitted from a subset. Note that these 32 subsets include T itself and the empty set, which contains no elements at all. If T contains all elements of concern, it is called the universal set. Given a subset of T, such as S, we may define the complement of S with respect to a universal set T to consist of precisely those elements of T which are not included in the given subset.

Thus, S as above defined has its complement (with respect to T) $\bar{S} = \{d, e\}$. The union of any two sets (subsets of a given set) consists of those elements that are in one or the other or in both given sets; the intersection of two sets consists of those elements that are in both given sets. We use the symbol $\cup$ to denote the union, and $\cap$ to denote the intersection of two sets. For example, if B={b, d, e}, then B$\cup$S={a, b, c, d, e}, and B$\cap$S={b}.

While other set operations may be defined, the operations of complementation union and intersection are of primary interest to us. A Boolean algebra is a finite or infinite set of elements together with three operations—negation, addition, and multiplication—that correspond to the set operations of complementation, union, and intersection, respectively. Among the elements of a Boolean algebra are two distinguished elements: 0, corresponding to the empty set; and 1, corresponding to the universal set. For any given element of a Boolean algebra, there is a unique complement a' with the property that a+a'=1 and aa'=0. Boolean addition and multiplication are associative and commutative, as are ordinary addition and multiplication, but otherwise have somewhat different properties. The principal properties are given in Table 3-1, where a, b, and c are any elements of a Boolean algebra.

Table 3-1

Distributivity	a(b+c)=ab+ac
	a+(bc)=(a+b)(a+c)
Idempotency	a+a=a
	aa=a
Absorption laws	a+ab=a
	a(a+b)=a
DeMorgan's laws	(a+b)'=a'b'
	(ab)'=a'+b'

Since a finite set of n elements has exactly 2^n subsets, and it can be shown that the finite Boolean algebras are precisely the finite set algebras, each finite Boolean algebra consists of exactly 2^n elements for some integer n. For example, the set algebra for the set T defined above corresponds to a Boolean algebra of 32 elements.

While it is possible to use a different symbol to denote each element of a Boolean algebra, it is often more useful to represent the 2^n elements of a finite Boolean algebra by binary vectors having n components. With such a representation the operations of the Boolean algebra are accomplished componentwise by considering each component as an independent two-element Boolean algebra. This corresponds to representing subsets of a finite set by binary vectors. For example, since the set T has five elements, we may represent its subsets by five-component binary vectors, each component denoting an element of the set T. A numeral 1 in the i-th component of the vector denotes the inclusion of the i-th element of that particular subset; a 0 denotes its exclusion. Thus, the subset S={a, b, c} has the binary vector representation {1,1,1,0,0}. The set operations become Boolean operations on the components of the vectors. This representation of sets, and the correspondence to Boolean or logical operations, is very useful in information retrieval. Because of it, sets of document and query characteristics may be easily and rapidly matched.

New Words & Expressions

Boolean algebra　布尔代数
set algebra　集合代数
empty set　空集
universal set　全集
subset　子集
complement　补
distributivity　分配律
absorption Law　吸收律
arise from　起因于，由……引出
componentwise　adv. 按照分量

intersection　交
finite set　有限集
infinite set　无限集
exclusion　"异"运算
associative　结合律
commutative　交换律
idempotency　同一律
correspond to　相应于
together with　和，加上
retrieval　n. 检索

Notes

1. Given a subset of T, such as S, we may define the complement of S with respect to a universal set T to consist of precisely those elements of T which are not included in the given subset. 给定T的一个子集，例如子集S，我们可以定义一个关于全集T的S的补集，其中正好包含那些不在子集S中而在T中的元素。

2. While other set operations may be defined, the operations of complementation, union, and intersection are of primary interest to us. 虽然我们可以定义一些其他集合运算，但求补、并和交运算是我们最感兴趣的三个集合运算。

数学公式的读法（Pronunciation of mathematical expressions）

下面给出大部分数学公式的读法。除非需要特别准确地说明，通常最短形式是首选读法。

一、逻辑（Logic）

∃　　　　there exists

∀   for all
p⇒q   p implies q / if p, then q
p⇔q   p if and only if q / p is equivalent to q / p and q are equivalent

二、集合（Sets）

x∈A   x belongs to A / x is an element (or a member) of A
x∉A   x does not belong to A / x is not an element (or a member) of A
A⊂B   A is contained in B / A is a subset of B
A⊃B   A contains B / B is a subset of A
A∩B   A cap B / A meet B/ A intersection B
A∪B   A cup B/ A join B / A union B
B/A   A minus B/the difference between A and B
A×B   A cross B / the Cartesian product of A and B(A 与 B 的笛卡尔积)

三、实数（Real numbers）

x+1   x plus one
x-1   x minus one
x±1   x plus or minus one
xy   xy / x multiplied by y
(x-y)(x+y)   x minus y, x plus y
$\dfrac{x}{y}$ x over y
=   the equals sign
x=5   x equals 5 / x is equal to 5
x≠5   x (is) not equal to 5
x≡y   x is equivalent to (or identical with) y
x>y   x is greater than y
x≥y   x is greater than or equal to y
x<y   x is less than y
x≤y   x is less than or equal to y
0<x<1   zero is less than x is less than 1
0≤x≤1   zero is less than or equal to x is less than or equal to 1
|x| mod x / modulus x
x^2 x squared / x (raised) to the power 2
x^3 x cubed
x^4 x to the fourth / x to the power four
x^n x to the nth / x to the power n
x^{-n} x to the (power) minus n

$\sqrt{x}$	(square) root x / the square root of x
$\sqrt[3]{x}$	cube root (of) x
$\sqrt[4]{x}$	fourth root (of) x
$\sqrt[n]{x}$	nth root (of) x
$(x+y)^2$	x plus y all squared
$\left(\dfrac{x}{y}\right)^2$	x over y all squared
$n!$	n factorial
$\hat{x}$	x hat
$\bar{x}$	x bar
$\tilde{x}$	x tilde
x_i	xi / x subscript i / x suffix i / x sub i
$\sum_{i=1}^{n} a_i$	the sum from i equals one to n a_i / the sum as i runs from 1 to n of the a_i

四、线性代数（Linear algebra）

$\|A\|$	the norm (or modulus) of x
$\overrightarrow{OA}$	OA / vector OA OA
$\overline{OA}$	OA / the length of the segment OA
A^T	A transpose / the transpose of A
A^{-1}	A inverse / the inverse of A

五、函数（Functions）

$f(x)$	fx / f of x / the function f of x
$f: S \to T$	a function f from S to T
$x \mapsto y$	x maps to y / x is sent (or mapped) to y
$f'(x)$	f prime x / f dash x / the (rst) derivative of f with respect to x
$f''(x)$	f double–prime x / f double–dash x / the second derivative of f with respect to x
$f'''(x)$	f triple–prime x / f triple–dash x / the third derivative of f with respect to x
$f^{(4)}$	four x / the fourth derivative of f with respect to x
$\dfrac{\partial f}{\partial x_1}$	the partial (derivative) of f with respect to x_1
$\dfrac{\partial^2 f}{\partial x_1^2}$	the second partial (derivative) of f with respect to x_1
$\int_0^{\infty}$	the integral from zero to infinity
$\lim\limits_{x \to 0}$	the limit as x approaches zero

$\lim\limits_{x \to +0}$	the limit as x approaches zero from above
$\lim\limits_{x \to -0}$	the limit as x approaches zero from below
$\log_e y$	log y to the base e / log to the base e of y / natural log (of) y
ln y	log y to the base e / log to the base e of y / natural log (of) y

需要注意的是，每个数学家对数学公式常常有各自的读法，在许多情况下，并不存在一个普遍接受的所谓"正确"读法。

一些在书写上有明确区别的公式在口头表达时经常难以区分，如 fx 的发音可以理解为以下任何一种：fx, f(x), f_x, FX, $\overline{FX}$, $\overrightarrow{FX}$。**这种差别通常通过上下文来区分**。只有在可能发生混淆、或要强调其观点时，数学家才使用较长的读法：f 乘以 x, 关于 x 的函数 f, f 下标 x, 直线 FX, 弧 FX 的长度, 矢量 FX.

同样，对于下列几对公式，数学家在演讲时不大可能表现出明显的区别，除了有时在音调或停顿长度上有一些差别外：

x+(y+z) and (x+y)+z, $\sqrt{ax+b}$ and $\sqrt{ax}+b$, $a^n - 1$ and a^{n-1}

Exercises

I. Answer the following questions

1. What is Leibniz's reasons for advocating the binary system?
2. Describe the relationship between Boolean algebras and set operation.
3. What are the principal properties of Boolean algebras?

Chapter 4 Elementary Data Structures

学习指导

堆栈和队列是两种典型的数据结构：在栈中，最近被插入的元素最先被删除，即栈是按后进先出（LIFO）的原则进行的；而队列是以先进先出（FIFO）的原则进行的。链表这个数据结构中的对象都以线性顺序排列的，但不同于数组的顺序是由数组下标确定，链表的顺序是由每个对象的指针确定。通过本章的学习，读者应掌握：
- 栈的特点，及入栈/出栈的操作；
- 队列的特点，及入队/出队的操作；
- 单链表、双向链表和循环链表的概念，及查询、插入和删除等操作的实现；
- 掌握常用英汉互译技巧。

4.1 Stacks and queues

Stacks and queues are dynamic sets in which the element removed from the set by the DELETE operation is prespecified. In a **stack**, the element deleted from the set is the one most recently inserted: the stack implements a **last-in, first-out**, or **LIFO**, policy. Similarly, in a **queue**, the element deleted is always the one that has been in the set for the longest time: the queue implements a **first-in, first out**, or **FIFO**, policy. There are several efficient ways to implement stacks and queues on a computer. In this section we show how to use a simple array to implement each.

4.1.1 Stacks

The INSERT operation on a stack is often called PUSH, and the DELETE operation, which does not take an element argument, is often called POP. These names are allusions to physical stacks, such as the spring-loaded stacks of plates used in cafeterias. The order in which plates are popped from the stack is the reverse of the order in which they were pushed onto the stack, since only the top plate is accessible.

As shown in Fig.4-1, we can implement a stack of at most n elements with an array $S[1,\_n]$. The array has an attribute $top[S]$ that indexes the most recently inserted element. The stack consists of elements $S[1\_top[S]]$, where $S[1]$ is the element at the bottom of the stack and $S[top[S]]$ is the element at the top.

top is element 17.

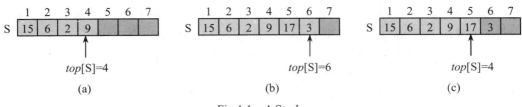

Fig.4-1   A Stack

Fig.4-1: An array implementation of a stack S. Stack elements appear only in the lightly shaded positions. *(a)* Stack S has 4 elements. The top element is 9. *(b)* Stack S after the calls PUSH(S, 17) and PUSH(S, 3). *(c)* Stack S after the call POP(S) has returned the element 3, which is the one most recently pushed. Although element 3 still appears in the array, it is no longer in the stack; the top is element 17.

When $top[S] = 0$, the stack contains no elements and is **empty**. The stack can be tested for emptiness by the query operation STACK-EMPTY. If an empty stack is popped, we say the stack **underflows**, which is normally an error. If $top[S]$ exceeds n, the stack **overflows**. (In our pseudocode implementation, we don't worry about stack overflow.)

The stack operations can each be implemented with a few lines of code.

STACK-EMPTY(S)
1 **if** $top[S] = 0$
2 **then return** TRUE
3 **else return** FALSE

PUSH(S, x)
1 $top[S] \leftarrow top[S] + 1$
2 $S[top[S]] \leftarrow x$

POP(S)
1 **if** STACK-EMPTY(S)
2 **then error** "underflow"
3 **else** $top[S] \leftarrow top[S] - 1$
4 **return** $S[top[S] + 1]$

Fig.4-1 shows the effects of the modifying operations PUSH and POP. Each of the three stack operations takes $O(1)$ time.

4.1.2   Queues

We call the INSERT operation on a queue ENQUEUE, and we call the DELETE operation DEQUEUE; like the stack operation POP, DEQUEUE takes no element argument. The FIFO property of a queue causes it to operate like a line of people in the registrar's office. The queue has a **head** and a **tail**. When an element is enqueued, it takes its place at the tail of the queue, just as a newly arriving student takes a place at the end of the line. The element dequeued is always the one at the head of the queue, like the student at the head of the line who has waited the longest. (Fortunately, we don't have to worry about computational elements cutting into line.)

Fig.4-2 shows one way to implement a queue of at most $n - 1$ elements using an array $Q[1\_n]$. The queue has an attribute $head[Q]$ that indexes, or points to, its head. The attribute $tail[Q]$ indexes the next location at which a newly arriving element will be inserted into the queue. The elements in the queue are in locations $head[Q]$, $head[Q] +1,..., tail[Q] - 1$, where we "wrap around" in the sense that location 1 immediately follows location n in a circularorder. When $head[Q] = tail[Q]$, the queue is empty. Initially, we have $head[Q] = tail[Q] = 1$.

When the queue is empty, an attempt to dequeue an element causes the queue to underflow. When $head[Q] = tail[Q] + 1$, the queue is full, and an attempt to enqueue an element causes the queue to overflow.

Fig.4-2: A queue implemented using an array $Q[1 \_ 12]$. Queue elements appear only in the lightly shaded positions. (a) The queue has 5 elements, in locations $Q[7 \_ 11]$. (b) The configuration of the queue after the calls ENQUEUE(Q, 17), ENQUEUE(Q, 3), and ENQUEUE(Q, 5). (c) The configuration of the queue after the call DEQUEUE(Q) returns the key value 15 formerly at the head of the queue. The new head has key 6.

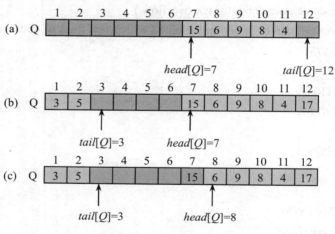

Fig.4-2 A Queue

In our procedures ENQUEUE and DEQUEUE, the error checking for underflow and overflow has been omitted.

ENQUEUE(Q, x)
1 $Q[tail[Q]] \leftarrow x$
2 **if** $tail[Q] = length[Q]$
3   **then** $tail[Q] \leftarrow 1$
4   **else** $tail[Q] \leftarrow tail[Q] + 1$
DEQUEUE(Q)
1 $x \leftarrow Q[head[Q]]$
2 **if** $head[Q] = length[Q]$
3   **then** $head[Q] \leftarrow 1$
4   **else** $head[Q] \leftarrow head[Q] + 1$

5 **return** x

Fig.4-2 shows the effects of the ENQUEUE and DEQUEUE operations. Each operation takes $O(1)$ time.

New Words & Expressions

stack   n. 栈，堆栈
attribute   n. 属性，性质
overflow   n. 上溢

queue   n. 队列
underflow   n. 下溢
pseudocode   n. 伪码

Abbreviations

LIFO（last in first out）后进先出   FIFO（fisrt in first out）先进先出

4.2   Linked lists

A **_linked list_** is a data structure in which the objects are arranged in a linear order. Unlike an array, though, in which the linear order is determined by the array indices, the order in a linked list is determined by a pointer in each object. Linked lists provide a simple, flexible representation for dynamic sets.

As shown in Fig.4-3, each element of a **_doubly linked list_** L is an object with a *key* field and two other pointer fields: *next* and *prev*. The object may also contain other satellite data. Given an element x in the list, *next*[x] points to its successor in the linked list, and *prev*[x] points to its predecessor. If *prev*[x] = NIL, the element x has no predecessor and is therefore the first element, or **_head_**, of the list. If *next*[x] = NIL, the element x has no successor and is therefore the last element, or **_tail_**, of the list. An attribute *head*[L] points to the first element of the list. If *head*[L] = NIL, the list is empty.

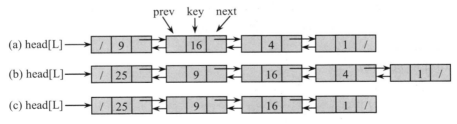

Fig.4-3   A doubly linked list

Fig.4-3: (a) A doubly linked list L representing the dynamic set {1, 4, 9, 16}. Each element in the list is an object with fields for the key and pointers (shown by arrows) to the next and previous objects. The *next* field of the tail and the *prev* field of the head are NIL, indicated by a diagonal slash. The attribute *head*[L] points to the head. (b) Following the execution of LIST-INSERT(L, x), where *key*[x] = 25, the linked list has a new object with key 25 as the new head. This new object points to the old head with key 9. (c) The result of the subsequent call LIST-DELETE(L, x), where x points to the object with key 4.

A list may have one of several forms. It may be either singly linked or doubly linked, it may be sorted or not, and it may be circular or not. If a list is **singly linked**, we omit the *prev* pointer in each element. If a list is **sorted**, the linear order of the list corresponds to the linear order of keys stored in elements of the list; the minimum element is the head of the list, and the maximum element is the tail. If the list is **unsorted**, the elements can appear in any order.

In a *circular list*, the *prev* pointer of the head of the list points to the tail, and the *next* pointer of the tail of the list points to the head. The list may thus be viewed as a ring of elements. In the remainder of this section, we assume that the lists with which we are working are unsorted and doubly linked.

4.2.1 Searching a linked list

The procedure LIST-SEARCH(L, k) finds the first element with key k in list L by a simple linear search, returning a pointer to this element. If no object with key k appears in the list, then NIL is returned. For the linked list in Fig.4-3(a), the call LIST-SEARCH(L, 4) returns a pointer to the third element, and the call LIST-SEARCH(L, 7) returns NIL.

LIST-SEARCH(L, k)
1 $x \leftarrow head[L]$
2 **while** $x \neq$ NIL and $key[x] \neq k$
3 **do** $x \leftarrow next[x]$
4 **return** x

To search a list of n objects, the LIST-SEARCH procedure takes $O(n)$ time in the worst case, since it may have to search the entire list.

4.2.2 Inserting into a linked list

Given an element x whose *key* field has already been set, the LIST-INSERT procedure "splices" x onto the front of the linked list, as shown in Fig.4-3(b).

LIST-INSERT(L, x)
1 $next[x] \leftarrow head[L]$
2 **if** $head[L] \neq$ NIL
3 **then** $prev[head[L]] \leftarrow x$
4 $head[L] \leftarrow x$
5 $prev[x] \leftarrow$ NIL

The running time for LIST-INSERT on a list of n elements is $O(1)$.

4.2.3 Deleting from a linked list

The procedure LIST-DELETE removes an element x from a linked list L. It must be given a pointer to x, and it then "splices" x out of the list by updating pointers. If we wish to delete an element with a given key, we must first call LIST-SEARCH to retrieve a pointer to the element.

LIST-DELETE(L, x)

1 **if** prev[x] ≠ NIL
2 **then** next[prev[x]] ← next[x]
3 **else** head[L] ← next[x]
4 **if** next[x] ≠ NIL
5 **then** prev[next[x]] ← prev[x]

Fig.4-3(c)shows how an element is deleted from a linked list. LIST-DELETE runs in $O(1)$ time, but if we wish to delete an element with a given key, $O(n)$ time is required in the worst case because we must first call LIST-SEARCH.

4.2.4 Sentinels

The code for LIST-DELETE would be simpler if we could ignore the boundary conditions at the head and tail of the list.

LIST-DELET' (L, x)
1 next[prev[x]] ← next[x]
2 prev[next[x]] ← prev[x]

A **sentinel** is a dummy object that allows us to simplify boundary conditions. For example, suppose that we provide with list L an object nil[L] that represents NIL but has all the fields of the other list elements. Wherever we have a reference to NIL in list code, we replace it by a reference to the sentinel nil[L]. As shown in Fig.4-4, this turns a regular doubly linked list into a **circular, doubly linked list with a sentinel**, in which the sentinel nil[L] is placed between the head and tail; the field next[nil[L]] points to the head of the list, and prev[nil[L]] points to the tail. Similarly, both the next field of the tail and the prev field of the head point to nil[L]. Since next[nil[L]] points to the head, we can eliminate the attribute head[L] altogether, replacing references to it by references to next[nil[L]]. An empty list consists of just the sentinel, since both next[nil[L]] and prev[nil[L]] can be set to nil[L].

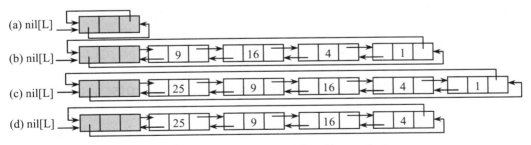

Fig.4-4 A circular, doubly linked list with a sentinel.

The sentinel nil[L] appears between the head and tail. The attribute head[L] is no longer needed, since we can access the head of the list by next[nil[L]]. *(a)* An empty list. *(b)* The linked list from Fig.4-3(a), with key 9 at the head and key 1 at the tail. *(c)* The list after executing LIST-INSER'(L, x), where key[x] = 25. The new object becomes the head of the list. *(d)* The list after deleting the object with key 1. The new tail is the object with key 4.

The code for LIST-SEARCH remains the same as before, but with the references to NIL and *head*[*L*] changed as specified above.

LIST-SEARC′(*L*, *k*)

1 $x \leftarrow next[nil[L]]$
2 **while** $x \neq nil[L]$ and $key[x] \neq k$
3 **do** $x \leftarrow next[x]$
4 **return** x

We use the two-line procedure LIST-DELET′ to delete an element from the list. We use the following procedure to insert an element into the list.

LIST-INSER′ (*L*, *x*)

1 $next[x] \leftarrow next[nil[L]]$
2 $prev[next[nil[L]]] \leftarrow x$
3 $next[nil[L]] \leftarrow x$
4 $prev[x] \leftarrow nil[L]$

Fig.4-4 shows the effects of LIST-INSER′ and LIST-DELET′ on a sample list. Sentinels rarely reduce the asymptotic time bounds of data structure operations, but they can reduce constant factors. The gain from using sentinels within loops is usually a matter of clarity of code rather than speed; the linked list code, for example, is simplified by the use of sentinels, but we save only $O(1)$ time in the LIST-INSER′ and LIST-DELET′ procedures. In other situations, however, the use of sentinels helps to tighten the code in a loop, thus reducing the coefficient of, say, n or $n2$ in the running time.

Sentinels should not be used indiscriminately. If there are many small lists, the extra storage used by their sentinels can represent significant wasted memory.

New Words & Expressions

linked list 链表	doubly linked list 双向链表
circular list 循环列表	diagonal slash 反斜线
retrieve v. 检索	sentinel n. 岗哨，哨兵位
asymptotic adj. 渐近的，渐近线的	coefficient n. 系数

Reading Material: Related Concepts

In this section we isolate three topics that are closely associated with the subject of data structures: abstraction, the distinction between static and dynamic structures, and the concept of a pointer.

Abstraction Again

The structures presented in the previous section are often associated with data. However, a computer's main memory is not organized as arrays, lists, stacks, queues, and trees but is instead organized as a sequence of addressable memory cells. Thus, all other structures must be simulated. How this simulation is accomplished is the subject of this chapter. For now we merely point out that organizations such as arrays, lists, stacks, queues, and trees are abstract tools that are created so that users of the data can be shielded from the details of actual data storage and can be allowed to access information as though it were stored in a more convenient form.

The term *user* in this context does not necessarily refer to a human. Instead, the meaning of the word depends on our perspective at the time. If we are thinking in terms of a person using a PC to maintain bowling league records, then the user is a human. In this case, the application software (perhaps a spreadsheet software package) would be responsible for presenting the data in an abstract form convenient to the human—most likely as a homogeneous array. If we are thinking in terms of a server on the Internet, then the user might be a client. In this case, the server would be responsible for presenting data in an abstract form convenient to the client. If we are thinking in terms of the modular structure of a program, then the user would be any module requiring access to the data. In this case, the module containing the data would be responsible for presenting the data in an abstract form convenient to the other modules. In each of these scenarios, the common thread is that the user has the privilege of accessing data as an abstract tool.

Static Versus Dynamic Structures

An important distinction in constructing abstract data structures is whether the structure being simulated is static or dynamic, that is, whether the shape or size of the structure changes over time. For example, if the abstract tool is a list of names, it is important to consider whether the list will remain a fixed size throughout its existence or expand and shrink as names are added and deleted.

As a general rule, static structures are more easily managed than dynamic ones. If a structure is static, we need merely to provide a means of accessing the various data items in the structure and perhaps a means of changing the values at designated locations. But, if the structure is dynamic, we must also deal with the problems of adding and deleting entries as well as finding the memory space required by a growing data structure. In the case of a poorly designed structure, adding a single new entry could result in a massive rearrangement of the structure, and excessive growth could dictate that the entire structure be transferred to another memory area where more space is available.

Pointers

Recall that the various cells in a machine's main memory are identified by numeric addresses. Being numeric values, these addresses themselves can be encoded and stored in memory cells. A **pointer** is a storage area that contains such an encoded address. In the case of data structures, pointers are used to record the location where data items are stored. For example, if we must repeatedly move an item of data from one location to another, we might designate a fixed location to serve as a pointer. Then, each time we move the item, we can update the pointer to reflect the new address of the data. Later, when we need to access the item of data, we can find it by means of the pointer. Indeed, the pointer will always "point" to the data. There we found that a register called a program counter is used to hold the address of the next instruction to be executed. Thus, the program counter plays the role of a pointer. In fact, another name for a program counter is **instruction pointer.**

As an example of the application of pointers, suppose we have a list of novels stored in a computer's memory alphabetically by title. Although convenient in many applications, this arrangement makes it difficult to find all the novels by a particular author—they are scattered throughout the list. To solve this problem, we can reserve an additional memory cell within each block of cells representing a novel and use this cell as a pointer to another block representing a book by the same author. In this manner the novels with common authorship can be linked in a loop. Once we find one novel by a given author, we can find all the others by following the pointers from one book to another.

Many modern programming languages include pointers as a primitive data type. That is, they allow the declaration, allocation, and manipulation of pointers in ways reminiscent of integers and character strings. Using such a language, a programmer can design elaborate networks of data within a machine's memory where pointers are used to link related items to each other.

New Words & Expressions

distinction　n. 区别，明显差别，特征　　　　　spreadsheet　n. 电子制表软件，电子数据表
homogeneous　adj. 同性质的，同类的　　　　　rearrangement　n. 重新整理
pointer　n. 指针

常用英汉互译技巧

英汉两种语言在句法、词汇、修辞等方面均存在着很大的差异，因此在进行英汉互译时必然会遇到很多困难，需要有一定的翻译技巧作指导。常用的翻译技巧有增译法、省译法、转换法、拆句法、合并法、正译法、反译法、倒置法、包孕法、插入法、重组法和综合法等。

一、增译法

指根据英汉两种语言不同的思维方式、语言习惯和表达方式，在翻译时增添一些词、短句或句子，以便更准确地表达出原文所包含的意义。这种方式多半用在汉译英里。

首先，汉语无主句较多，而英语句子一般都要有主语，所以在翻译汉语无主句的时候，除了少数可用英语无主句、被动语态或"There be…"结构来翻译以外，一般都要根据语境补出主语，使句子完整。

第二，英汉两种语言在名词、代词、连词、介词和冠词的使用方法上也存在很大差别。英语中代词使用频率较高，凡说到人的器官和归某人所有的或与某人有关的事物时，必须在前面加上物主代词。因此，在汉译英时需要增补物主代词，而在英译汉时又需要根据情况适当地删减。

其次，英语词与词、词组与词组以及句子与句子的逻辑关系一般用连词来表示，而汉语则往往通过上下文和语序来表示这种关系。因此，在汉译英时常常需要增补连词。英语句子离不开介词和冠词。

另外，在汉译英时还要注意增补一些原文中暗含而没有明言的词语和一些概括性、注释性的词语，以确保译文意思的完整。总之，通过增译，一是保证译文语法结构的完整，二是保证译文意思的明确。如：

例 1. Indeed, the reverse is true

实际情况恰好相反。（增译名词）

例 2. 这是这两代计算机之间的又一个共同点。

This is yet another common point between the computers of the two generations.（增译介词）

例 3. Individual mathematicians often have their own way of pronouncing mathematical expressions and in many cases there is no generally accepted "correct" pronunciation.

每个数学家对数学公式常常有各自的读法，在许多情况下，并不存在一个普遍接受的所谓"正确"读法。（增加隐含意义的词）

例 4. 只有在可能发生混淆、或要强调其观点时，数学家才使用较长的读法

It is only when confusion may occur, or where he/she wishes to emphasis the point, that the mathematician will use the longer forms.（增加主语）

例 5. 三个臭皮匠，合成一个诸葛亮。

Three cobblers with their wits combined equal Zhuge Liang the mastermind.（增译注释性词语）

例 6. Three computers in this section are infected by virus.

这个部门有三台计算机感染了病毒。（增译量词）

二、省译法

这是与增译法相对应的一种翻译方法，即删去不符合目标语思维习惯、语言习惯和表达方式的词，以避免译文累赘。增译法的例句反之即可。又如：

例 1. You will be staying in this hotel during your visit in Beijing.

你在北京访问期间就住在这家饭店里。（省译物主代词）

例 2. I hope you will enjoy your stay here.

希望您在这儿过得愉快。（省译主语）

例 3. 中国政府历来重视环境保护工作。

The Chinese government has always attached great importance to environmental protection.（省译名词）

例 4. The development of IC made it possible for electronic devices to become smaller and smaller.

集成电路的发展是电子器件可以做得越来越小。（省译形式主语 it）

三、转换法

转换法指在翻译过程中为了使译文符合目标语的表述方式、方法和习惯而对原句中的词类、句型和语态等进行转换。具体地说，就是在词性方面，把名词转换为代词、形容词、动词；把动词转换成名词、形容词、副词、介词；把形容词转换成副词和短语。在句子成分方面，把主语变成状语、定语、宾语、表语；把谓语变成主语、定语、表语；把定语变成状语、主语；把宾语变成主语。在句型方面，把并列句变成复合句，把复合句变成并列句，把状语从句变成定语从句。在语态方面，可以把主动语态变为被动语态。如：

例 1. Too much exposure to TV programs will do great harm to the eyesight of children.

孩子们看电视过多会大大地损坏视力。（名词转动词）

例 2. 由于我们实行了改革开放政策，我国的综合国力有了明显的增强。

Thanks to the introduction of our reform and opening policy, our comprehensive national strength has greatly improved.（动词转名词）

例 3. In his article the author is critical of man's negligence toward his environment.

作者在文章中，对人类疏忽自身环境作了批评。（形容词转名词）

例 4. 时间不早了，我们回去吧！

We don't have much time left. Let's go back.（句型转换）

例 5. 学生们都应该德、智、体全面发展。

All the students should develop morally, intellectually and physically.（名词转副词）

例 6. The modern world is experiencing rapid development of information technology.

当今世界的信息技术正在迅速发展。（名词转动词、形容词转副词）

四、拆句法和合并法

这是两种相对应的翻译方法。拆句法是把一个长而复杂的句子拆译成若干个较短、较简单的句子，通常用于英译汉；合并法是把若干个短句合并成一个长句，一般用于汉译英。汉语强调意合，结构较松散，因此简单句较多；英语强调形合，结构较严密，因此长句较多。所以汉译英时要根据需要注意利用连词、分词、介词、不定式、定语从句、独立结构等把汉语短句连成长句；而英译汉时又常常要在原句的关系代词、关系副词、主谓连接处、并列或转折连接处、后续成分与主体的连接处，以及意群结束处将长句切断，译成汉语分句。这样就可以基本保留英语语序，顺译全句，顺应现代汉语长短句相替、单复句相间的句法修辞原则。如：

例 1. Increased cooperation with China is in the interests of the United States.

同中国加强合作，符合美国的利益。（在主谓连接处拆译）

例 2. It is common practice the electric wires are made from copper.

电线是铜制成的。（主从句合一）

例 3. 中国是个大国，百分之八十的人口从事农业，但耕地只占土地面积的十分之一，其余为山脉、森林、城镇和其他用地。

China is a large country with four-fifths of the population engaged in agriculture, but only one tenth of the land is farmland, the rest being mountains, forests and places for urban and other uses.（合译法）

例 4. Packet switching is a method of slicing digital messages into parcels called "packets," sending the packets along different communication paths as they become available, and then reassembling the packets once they arrive at their destination.

分组交换是传输数据的一种方法，它先将数据信息分割成许多称为"分组"的数据信息包；当路径可用时，经过不同的通信路径发送；当到达目的地后，再将它们组装起来。（将长定语从句拆成几个并列的分句）

例 5. The URL htttp://www.azimuth-interactive.com/flash_test refers to the IP address 208.148.84.1 with the domain name "azimuth-interactive.com" and the protocol being used to access the address, Hypertext Transfer Protocol (HTTP).

网址 htttp://www.azimuth-interactive.com/flash_test 就是指 IP 地址 208.148.84.1，其域名为

"azimuth-interactive.com",而访问地址时使用的协议为超文本传输协议 HTTP。(拆分法)

五、正译法和反译法

这两种方法通常用于汉译英,偶尔也用于英译汉。所谓正译,是指把句子按照与汉语相同的语序或表达方式译成英语。所谓反译则是指把句子按照与汉语相反的语序或表达方式译成英语。正译与反译常常具有同义的效果,但反译往往更符合英语的思维方式和表达习惯。因此比较地道。如:

例1. 在美国,人人都能买到枪。
In the United States, everyone can buy a gun.(正译)
In the United States, guns are available to everyone.(反译)

例2. 你可以从因特网上获得这一信息。
You can obtain this information on the Internet.(正译)
This information is accessible/available on the Internet.(反译)

例3. 他突然想到了一个新主意。
Suddenly he had a new idea.(正译)
He suddenly thought out a new idea.(正译)
A new idea suddenly occurred to/struck him.(反译)

例4. 他仍然没有弄懂我的意思。
He still could not understand me.(正译)
Still he failed to understand me.(反译)

例5. 无论如何,她算不上一位思维敏捷的学生。
She can hardly be rated as a bright student.(正译)
She is anything but a bright student.(反译)

例6. Please withhold the document for the time being.
请暂时扣下这份文件。(正译)
请暂时不要发这份文件。(反译)

六、倒置法

在汉语中,定语修饰语和状语修饰语往往位于被修饰语之前;在英语中,许多修饰语常常位于被修饰语之后,因此翻译时往往要把原文的语序颠倒过来。倒置法通常用于英译汉,即对英语长句按照汉语的习惯表达法进行前后调换,按意群或进行全部倒置,原则是使汉语译句安排符合现代汉语论理叙事的一般逻辑顺序。有时倒置法也用于汉译英。如:

例1. At this moment, through the wonder of telecommunications, more people are seeing and hearing what we say than on any other occasions in the whole history of the world.
此时此刻,通过现代通信手段的奇迹,看到和听到我们讲话的人比整个世界历史上任何其他这样的场合都要多。(部分倒置)

例2. 改革开放以来,中国发生了巨大的变化。
Great changes have taken place in China since the introduction of the reform and opening policy.(全部倒置)

七、包孕法

这种方法多用于英译汉。所谓包孕是指在把英语长句译成汉语时,把英语后置成分按照汉语的正常语序放在中心词之前,使修饰成分在汉语句中形成前置包孕。但修饰成分不宜过长,否则会形成拖沓或造成汉语句子成分在连接上的纠葛。如:

例1. **IP multicasting** is a set of technologies that enables efficient delivery of data to many locations on a network.

IP 多信道广播是使数据向网络中许多位置高效传送的一组技术。

例2. What brings us together is that we have common interests which transcend those differences.

使我们走到一起的,是我们有超越这些分歧的共同利益。

八、插入法

指把难以处理的句子成分用破折号、括号或前后逗号插入译句中。这种方法主要用于笔译中。偶尔也用于口译中,即用同位语、插入语或定语从句来处理一些解释性成分。如:

如果说宣布收回香港就会像夫人说的"带来灾难性的影响",那我们将勇敢地面对这个灾难,做出决策。

If the announcement of the recovery of Hong Kong would bring about, as Madam put it, "disastrous effects," we will face that disaster squarely and make a new policy decision.

九、重组法

指在进行英译汉时,为了使译文流畅和更符合汉语叙事论理的习惯,在捋清英语长句的结构、弄懂英语原意的基础上,彻底摆脱原文语序和句子形式,对句子进行重新组合。如:

Decision must be made very rapidly; physical endurance is tested as much as perception, because an enormous amount of time must be spent making certain that the key figures act on the basis of the same information and purpose.

必须把大量时间花在确保关键人物均根据同一情报和目的行事,而这一切对身体的耐力和思维能力都是一大考验。因此,一旦考虑成熟,决策者就应迅速做出决策。

十、综合法

是指单用某种翻译技巧无法译出时,着眼篇章,以逻辑分析为基础,同时使用转换法、倒置法、增译法、省译法、拆句法等多种翻译技巧的方法。如:

例1. Behind this formal definition are three extremely important concepts that are the basis for understanding the Internet: packet switching, the TCP/IP communications protocol, and client/server computing.

在这个正式的定义背后,隐含着三个极其重要的概念:分组交换、TCP/IP(传输控制协议/网际协议)通信协议和客户机/服务器计算技术,它们乃是理解因特网的基础。

例2. Routers are special purpose computers that interconnect the thousands of different computer networks that make up the Internet and route packets along to their ultimate destination as they travel.

路由器是一种特殊用途的计算机，它将组成因特网的成千上万个不同计算机网络互相联接起来，并在信息包旅行时将它们向终极目的地发送。

Exercises

I. Explain the following terms in your own words

stack   queue   linked list

II. Fill in the following blanks

1. The process of inserting a element on the stack is called a _____ operation, and the process of deleting a element is called a _____ operation.
2. We call the INSERT operation on a queue _____, and we call the DELETE operation _____.
3. A queue is a dynamic set that obeys the _____ property.
4. A _____ is one of the most fundamental data structures, the order in which is determined by a pointer in each object.
5. Each element of a doubly linked list is an object with a _____ field and two other pointer fields.

III. List four types of linked list:

1. _____
2. _____
3. _____
4. _____

Chapter 5　Operating System

学习指导

操作系统从本质上讲是一个软件集合，是配置在计算机硬件上的第一层软件，是对计算机功能的第一次扩充。它负责计算机全部软、硬件资源的分配工作，实现信息的存取和保护，给用户提供接口，使用户获得良好的工作环境。通过本章的学习，读者应掌握：
- 操作系统的基本功能；
- 几种服务器用操作系统的性能比较；
- 被动语态的翻译技巧。

An Operating System or OS is a software program that enables the computer hardware to communicate and operate with the computer software. Without a computer Operating System a computer would be useless.

5.1　OS Functions

OS functions can be classified into
- Resource allocation and related functions.
- User interface functions.

The resource allocation function implements resource sharing by the users of a computer system. Basically, it performs binding of a set of resources with a requesting program, that is, it associates resources with a program. The related functions implement protection of users sharing a set of resources against mutual interference.

The user interface function facilitates creation and use of appropriate computational structures by a user. This function typically involves the use of a command language or a menu.

5.1.1　Resource allocation and related functions

The resource allocation function allocates resources for use by a user's computation. Resources can be divided into system provided resources like CPUs, memory areas and IO devices, or user-created resources like files which are entrusted to the OS.

Resource allocation criteria depend on whether a resource is a system resource or a user-created resource. Allocation of system resources is driven by considerations of efficiency of resource utilization. Allocation of user-created resources is based on a set of constraints specified by its creator and typically embodies the notion of access privileges.

Two popular strategies for resource allocation are:
- Partitioning of resources.
- Allocation from a pool.

In the resource partitioning approach, the OS decides apriori what resources should be allocated to a user computation. This approach is called static allocation because the allocation is made before the execution of a program starts. Static resource allocation is simple to implement, however, it could lead to suboptimal utilization because the allocation is made on the basis of perceived needs of a program, rather than its actual needs. In the latter approach, the OS maintains a common pool of resources and allocates from this pool on a need basis. Thus, OS considers allocation of a resource when a program raises a request for a resource. This approach is called dynamic allocation because the allocation takes place during the execution of a program. Dynamic resource allocation can lead to better utilization of resources because the allocation is made when a program requests a resource.

An OS can use a resource table as the central data structure allocation. The table contains an entry for each resource unit in the system. The entry contains the name or address of the resource unit and its present status, i.e. whether it is free or allocated to some program. When a program raises a request for a resource, the resource would be allocated to it if it is presently free. If many resource units of a resource class exist in the system, a resource request only indicates the resource class and the OS checks if any resource unit of that class is available for allocation.

In the partitioned resource allocation approach, the OS decides on the resources to be allocated to a program based on the number of resources and the number of programs in the system. For example, an OS may decide that a program can be allocated 1 MB of memory, 2000 disk blocks and a monitor. Such a collection of resources is referred to as a partition. In effect, a set of partitions can be predefined in the system. The resource table can have an entry for each resource partition. When a new program is to be started, an available partition is allocated to it.

Resource preemption

There are different ways in which resource can be shared by a set of programs. Some of these are:
- Sequential sharing.
- Concurrent sharing.

In sequential sharing, a resource is allocated for exclusive use by a program. When the resource is de-allocated, it is marked free in the resource table. Now it can be allocated to another program. In concurrent sharing, two or more programs can concurrently use the same resource. Examples of concurrent sharing are data files. Most other resources cannot be shared concurrently. Unless otherwise mentioned, all through this text resources are assumed to be only sequentially shareable.

When a resource is sequentially shareable, the system can de-allocate a resource when the program makes an explicit request for de-allocation. Alternatively, it can de-allocate a resource by force. This is called resource preemption, that is, forceful de-allocation of a resource.

Preemption of system resources is used by the OS to enforce fairness in their use by programs, or to realize certain system level goals. A preempted program cannot execute unless the preempted resource unit, or some other resource unit pi' the same resource class, is allocated to it once again.

The shorter term preemption is used preemption of the CPU, and the full term resource preemption is used for preemption of other resources.

CPU sharing

The CPU can be shared in a sequential manner only. Hence only one program can execute atany time. Other programs in the system have to wait their turn. It is often important to provide fairservice to all programs in the system. Hence preemption is used to free the CPU so that it can begiven to another program. Deciding which program should be given the CPU and for how long isa critical function. This function is called CPU Scheduling, or simply scheduling. Partitioning is abad approach for CPU Sharing, allocation from a pool is the obvious approach to use.

Memory sharing

Like the CPU, the memory also cannot be shared concurrently. However, unlike the CPU, its availability can be increased by treating different parts of memory as different resources. Both the partitioning and the pool-based allocation approaches can be used to manage the memory resource. Memory preemption can also be used to increase the availability of memory to programs. Special terms are used for different memory preemption techniques, hence the term "memory preemption" is rarely used in our discussions.

5.1.2 User interface related functions

The purpose of a user interface is to provide for the use of OS resources, primarily the CPU, for processing a user's computational requirements. OS user interfaces typically use command languages. The user uses a command to set up an appropriate computational structure to fulfill a computational requirement.

A variety of computational structures can be defined by an OS. A sample list of computational structures is as follows:

1. A single program.
2. A sequence of single programs.
3. A collection of programs.

These computational structures will be defined and described in later chapters; here we only point out a few salient features. It is assumed that each program is individually initiated by the user through the user interface. The single program consists of the execution of a program on a given set of data. The user initiates execution of the program through a command. Two kinds of programs can exist sequential and concurrent. A sequential program, which matches with the conventional notion of a program, is the simplest computational structure. In a concurrent program, different parts of the program can execute concurrently. For this, the OS has to be aware the identities of the different parts which can execute concurrently. This function is typically not served by the user interface of the OS. In this chapter it is assumed that each program is sequential in nature.

In a sequence of programs, each program is initiated by the user individually. However, the programs are not independent of each other——executing of a program is meaningful only if the previous programs in the sequence execute successfully. However, since the programs are initiated individually, their interface with one another is set up explicitly by the user. In a collection of programs, the user names the programs involved in the collection in his command. Thus, their identities are indicated to the OS through the user interface itself. The interface between the programs is handled by the OS.

New Words and Expressions

allocation   n. 分配，安置
binding   n. 捆绑，绑定
entrust   v. 委托
constraint   n. 约束，强制，局促
apriori   预先，事前
suboptimal   adj. 未达最佳标准的
sequential   adj. 顺序的，串行的
exclusive use   专用
scheduling   n. 调度

interface   n. 界面，接口
mutual interference   相互干扰
criteria   n. 标准
partition   vt. 分区
perceive   v. 感知，感到，认识到
preemption   n. 抢占
concurrent   adj. 并发的，并行的
de-allocate   vt. 释放
salient   adj. 重要的，显著的

Abbreviations

OS (Operating System)   操作系统

5.2 FreeBSD vs. Linux vs. Windows 2000

Reliability

FreeBSD is extremely robust. There are numerous testimonials of active servers with uptimes measured in years. The new Soft Updates file system optimizes disk I/O for high performance, yet still ensures reliability for transaction based applications, such as databases.

Linux is well known for its reliability. Servers often stay up for years. However, disk I/O is non-synchronous by default, which is less reliable for transaction based operations, and can produce a corrupted file system after a system crash or power failure. But for the average user, Linux is a very dependable OS.

All Windows users are familiar with the "Blue Screen of Death". Poor reliability is one of the major drawbacks of Windows. Some of the major issues have been fixed in Windows 2000, but "code bloat" has introduced many more reliability problems. Windows 2000 uses a lot of system resources and it is very difficult to keep the system up for more than a couple of months without it reverting to a crawl as memory gets corrupted and file systems fragmented.

Performance

FreeBSD is the system of choice for high performance network applications. FreeBSD will outperform other systems when running on equivalent hardware. The largest and busiest public server on the Internet, at ftp://ftp.cdrom.com/, uses FreeBSD to serve more than 1.2TB/day of downloads. FreeBSD is used by **Yahoo!, Qwest** and many others as their main server OS because of its ability to handle heavy network traffic with high performance and rock solid reliability.

Linux performs well for most applications, however the performance is not optimal under heavy network load. The network performance of Linux is 20-30% below the capacity of FreeBSD running on the same hardware. The situation has improved somewhat recently and the 2.4 release of the Linux kernel will introduce a new virutual memory system based on the same concepts as the FreeBSD VM system. Since both operating systems are open source, beneficial technologies are shared and for this reason the performance of Linux and FreeBSD is rapidly converging.

Windows is adequate for routine desktop apps, but it is unable to handle heavy network loads. A few organizations try to make it work as an Internet server. For instance, barnesandnoble.com uses Windows-NT, as can be verified by the error messages that their webserver produces, such as this recent example: **Error Message: [Microsoft][ODBC SQL Server Driver][SQL Server]Can't allocate space for object 'queryHistory' in database 'web' because the 'default' segment is full.**

For their own "Hotmail" Internet servers, Microsoft used FreeBSD for many years.

Security

FreeBSD has been the subject of a massive auditing project for several years. All of the critical system components have been checked and rechecked for security-related errors. The entire system is open source so the security of the system can and has been verified by third parties. A default FreeBSD installation has yet to be affected by a single CERT security advisory in 2000.

FreeBSD also has the notion of kernel security levels. These are much more powerful than simple run-levels since they allow the administrator to completely deny access to certain operating system functions such as reading /dev/mem, changing file system flags, or writing to disks without mounting a file system. FreeBSD includes a very robust packet filtering firewall system and many intrusion detection tools.

The open source nature of Linux allows anyone to inspect the security of the code and make changes, but in reality the Linux codebase is modified too rapidly by inexperienced programmers. There is no formal code review policy and for this reason Linux has been susceptible to nearly every Unix-based CERT advisory of the year.

However, Linux does include a very robust packet filtering firewall system and many intrusion detection tools.

Microsoft claims that their products are secure. But they offer no guarantee, and their software is not available for inspection or peer review. Since Windows is *closed source* there is no way for users to fix or diagnose any of the security compromises that are regularly published about Microsoft systems.

Device Drivers

The FreeBSD bootloader can load binary drivers at boot-time. This allows third-party driver manufacturers to distribute binary-only driver modules that can be loaded into any FreeBSD system. Due to the open-source nature of FreeBSD, it is very easy to develop device drivers for new hardware. Unfortunately, most device-manufacturers will only release binaries for Microsoft operating systems. This means that it can take several months after a hardware device has hit the market until a device driver is available.

The Linux community intentionally makes it difficult for hardware manufacturers to release binary-only drivers. This is meant to encourage hardware manufactureres to develop open-source device drivers. Unfortunately most vendors have been unwilling to release the source for their drivers so it is very difficult for Linux users to use vendor supplied drivers at all.

Microsoft has excellent relationships with hardware vendors. There are often conflicts when using a device driver on different versions of Microsoft Windows, but overall Windows users have excellent access to third party device drivers.

Commercial Applications

The number of commercial applications for FreeBSD is growing rapidly, but is still below what is available for Windows. In addition to native applications, FreeBSD can also run programs compiled for Linux, SCO Unix, and BSD/OS.

Many new commercial applications are available for Linux, and more are being developed. Unfortunately, Linux can only run binaries that are specifically compiled for Linux. It is unable to run programs compiled for FreeBSD, SCO Unix, or other popular operating systems.

There are thousands of applications available for Windows, far more than for any other OS. Nearly all commercial desktop applications run on Windows, and many of them are only available on Windows. If you have an important application that only runs on Windows, then you may have no choice but to run Microsoft Windows.

Development environment

FreeBSD includes an extensive collection of development tools. You get a complete C/C++ development system (editor, compiler, debugger, profiler, etc.) and powerful Unix development tools for Java, HTTP, Perl, Python, Tcl/Tk, Awk, Sed, etc. All of these are free, and are included in the basic FreeBSD installation. All come with full source code.

Linux includes all the same development tools as FreeBSD, with compilers and interpreters for every common programming language, all the GNU programs, including the powerful GNU C/C++ Compiler, Emacs editor, and GDB debugger. Unfortunately due to the very splintered nature of Linux, applications that you compile on one system (Red Hat 7) may not work on another Linux system (Slackware).

Very few development tools are included with Windows 2000. Most need to be purchased separately, and are rarely compatible with each other.

Support

FreeBSD can be downloaded from the Internet for FREE. Or it can be purchased on a four CDROM set, along with several gigabytes of applications, for $40. All necessary documentation is included. Support is available for free or for very low cost. There is no user licensing, so you can quickly bring additional computers online. This all adds up to a very low total cost of Ownership.

Linux is FREE. Several companies offer commercial aggregations at a very low cost. Applications and Documentation is available for little or no cost. There are no licensing restrictions, so Linux can be installed on as many systems as you like for no additional cost. Linux's total cost of ownership is very low.

The server edition of Windows 2000 costs nearly $700. Even basic applications cost extra. Users often spend many thousands of dollars for programs that are included for free with Linux or FreeBSD. Documentation is expensive, and very little on-line documentation is provided. A license is required for every computer, which means delays and administrative overhead. The initial learning curve for simple administration tasks is smaller than with Unix, but it also requires a lot more work to keep the system running with any significant work load.

New Words and Expressions

reliability   n. 可靠性
security-related   adj. 与安全相关的
audit   v. 审计，查核
intrusion   n. 入侵

Reading Material: RATs

The world of malicious software is often divided into two types: viral and nonviral. Viruses are little bits of code that are buried in other codes. When the "host" codes are executed, the viruses replicate themselves and may attempt to do something destructive. In this, they behave much like biological viruses.

Worms are a kind of computer parasite considered to be part of the viral camp because they replicate and spread from computer to computer.

As with viruses, a worm's malicious act is often the very act of replication; they can overwhelm computer infrastructures by generating massive numbers of e-mails or requests for connections that servers can't handle.

Worms differ from viruses, though, in that they aren't just bits of code that exist in other files. They could be whole files ——an entire Excel spreadsheet, for example. They replicate without the need for another program to be run.

Remote administration types are an example of another kind of nonviral malicious software, the Trojan horse, or more simply Trojan. The purpose of these programs isn't replication, but to penetrate and control. That masquerade as one thing when in fact they are something else, usually something destructive.

There are a number of kinds of Trojans, including spybots, which report on the web sites a computer user visits, and keybots or keyloggers, which record and report the user's keystrokes in order to discover passwords and other confidential information.

RAT's attempt to five a remote intruder administrative control of an infected computer. They work as client/server pairs. The server resides on the infected machine, while the client resides elsewhere, across the network, where it's available to a remote intruder.

Using standard TCP/IP or UDP protocols, the client sends instructions to the server. The server does what it's told to do on the infected computer.

Trojans, including RATs, are usually downloaded inadvertently by even the most savvy users. Visiting the wrong web site or clicking on the wrong hyperlink invites the unwanted Trojan in. RATs install themselves by exploiting weaknesses in standard programs and browsers.

Once they reside on a computer, RATs are hard to detect and remove. For Windows users, simply pressing Ctrl-Alt-Delete won't expose RATs, because they operate in the background and don't appear in the task list.

Some especially nefarious RATs have been designed to install themselves in such a way that they're very difficult to remove even after they're discovered.

For example, a variant of the Back Orifice RAT called G_Door installs its server as Kernel32.exe in the Windows system directory, where it's active and locked and controls the registry keys.

The active Kernel32.exe can't be removed, and a reboot won't clear the registry keys. Every time an infected computer stars, Kernel32.exe will be restarted, and the program will be active and locked.

Some RAT servers listen on known or standard ports. Others listen on random ports, telling their clients which port and which IP address to connect to by e-mail.

Even computer that connect to the Internet through Internet service providers, which are often thought to offer better security than static broadband connections, can be susceptible to control from such RAT servers.

The ability of RAT servers to initiate connections can also allow some of them to evade firewalls. An outgoing connection is usually permitted. Once a server contacts a client, the client and server can communicate, and the server begins following the instructions of the client.

legitimate tools are used by systems administrators to manage networks for a variety of reasons, such as logging employee usage and downloading program upgrades ——functions that are remarkably similar to those of some remote administration Trojans. The distinction between the two can be quite narrow. A remote administration tool used by an intruder becomes a RAT.

In April 2001, an unemployed British systems administrator named Gary McKinnon used a legitimate remote administration tool known as Remotely Anywhere to gain control of computers on a U.S. Navy network.

By hacking a few unguarded passwords on the target computers and using illegal copies of Remote Anywhere, McKinnon launched the attack from his girlfriend's e-mail account left him vulnerable to detection.

Some of the famous RATs are variants of Back Orifice; they include Netbus, SubSeven, Bionet and Hack'a'tack. These RATs tend to be families more than single programs. They are morphed by hackers into a vast array of Trojans with similar capabilities.

New Words & Expressions

malicious	adj. 恶意的	viral	adj. 病毒性的
replicate	v. 复制	parasite	n. 寄生虫
penetrate	v. 渗透	masquerade	v. 伪装
spybot	n. 间谍机器人	keybot	n. 击键机器人

intruder　　　　n. 入侵者　　　　　　　morph　　　　v. 变形，变体

Abbreviations

RAT(Remote Administration Trojan)　远程管理特洛伊木马（病毒）

被动语态的译法

英语中被动语态的使用范围极为广泛，尤其是在科技英语中，被动语态几乎随处可见，凡是在不必、不愿说出或不知道主动者的情况下均可使用被动语态，因此，掌握被动语态的翻译方法是极为重要的。在汉语中，也有被动语态，通常通过"把"或"被"等词体现出来，但它的使用范围远远小于英语中被动语态的使用范围，因此英语中的被动语态在很多情况下都翻译成主动结构。

对于英语原文的被动结构，我们一般采取下列的方法：

一、翻译成汉语的主动句

英语原文的被动结构翻译成汉语的主动结构又可以进一步分为几种不同的情况。

1.1 英语原文中的主语在译文中仍做主语

在采用此方法时，我们往往在译文中使用"加以""经过""用……来"等词来体现原文中的被动含义。例如：

例1. Other questions will be discussed briefly.

其他问题将简单地加以讨论。

例2. In other words mineral substances which are found on earth must be extracted by digging, boring holes, artificial explosions, or similar operations which make them available to us.

换言之，矿物就是存在于地球上，但须经过挖掘、钻孔、人工爆破或类似作业才能获得的物质。

例3. Nuclear power's danger to health, safety, and even life itself can be summed up in one word: radiation.

核能对健康、安全，甚至对生命本身构成的危险可以用一个词——辐射来概括。

1.2 将英语原文中的主语翻译为宾语，同时增补泛指性的词语（人们，大家等）做主语

例1. It could be argued that the radio performs this service as well, but on television everything is much more living, much more real.

可能有人会指出，无线电广播同样也能做到这一点，但还是电视屏幕上的节目要生动、真实得多。

例2. Television, it is often said, keeps one informed about current events, allows one to follow the latest developments in science and politics, and offers an endless series of programs which are both instructive and entertaining.

人们常说，电视使人了解时事，熟悉政治领域的最新发展变化，并能源源不断地为观众提供各种既有教育意义又有趣的节目。

例3. It is generally accepted that the experiences of the child in his first years largely determine his character and later personality.

人们普遍认为，孩子们的早年经历在很大程度上决定了他们的性格及其未来的人品。

另外，下列的结构也可以通过这一手段翻译：

It is asserted that … 有人主张 ……

It is believed that … 有人认为……

It is generally considered that … 大家（一般人）认为

It is well known that … 大家知道（众所周知）……

It will be said … 有人会说……

It was told that … 有人曾经说……

1.3 将英语原文中的 by, in, for 等做状语的介词短语翻译成译文的主语，而英语原文中的主语一般被翻译成宾语

例 1. A right kind of fuel is needed for an atomic reactor.

原子反应堆需要一种合适的燃料。

例 2. By the end of the war, 800 people had been saved by the organization, but at a cost of 200 Belgian and French lives.

大战结束时，这个组织拯救了八百人，但那是以二百多比利时人和法国人的生命为代价的。

例 3. And it is imagined by many that the operations of the common mind can be by no means compared with these processes, and that they have to be acquired by a sort of special training.

许多人认为，普通人的思维活动根本无法与科学家的思维过程相比，而且认为这些思维过程必须经过某种专门的训练才能掌握。

1.4 翻译成汉语的无主句

例 1. Great efforts should be made to inform young people especially the dreadful consequences of taking up the habit.

应该尽最大努力告诫年轻人吸烟的危害，特别是吸上烟瘾后的可怕后果。

例 2. By this procedure, different honeys have been found to vary widely in the sensitivity of their inhibit to heat.

通过这种方法分析发现不同种类的蜂蜜的抗菌活动对热的敏感程度也极为不同。

例 3. Many strange new means of transport have been developed in our century, the strangest of them being perhaps the hovercraft.

在我们这个世纪内研制了许多新奇的交通工具，其中最奇特的也许就是气垫船了。

例 4. New source of energy must be found, and this will take time….

必须找到新的能源，这需要时间……

另外，下列结构也可以通过这一手段翻译：

It is hoped that … 希望……

It is reported that … 据报道……

It is said that … 据说……

It is supposed that … 据推测……

It may be said without fear of exaggeration that … 可以毫不夸张地说……

It must be admitted that … 必须承认……

It must be pointed out that … 必须指出……
It will be seen from this that … 由此可见……

1.5 翻译成带表语的主动句

例1. The decision to attack was not taken lightly.
进攻的决定不是轻易做出的。

例2. On the whole such an conclusion can be drawn with a certain degree of confidence, but only if the child can be assumed to have had the same attitude towards the test as the other with whom he is being compared, and only if he was not punished by lack of relevant information which they possessed.
总的来说，得出这种结论是有一定程度把握的，但必须具备两个条件：能够假定这个孩子对测试的态度和与他比较的另一个孩子的态度相同；他也没有因为缺乏别的孩子已掌握的有关知识而被扣分。

二、译成汉语的被动语态

英语中的许多被动句可以翻译成汉语的被动句。常用"被""给""遭""挨""为……所""使""由…""受到"等表示。例如：

例1. Early fires on the earth were certainly caused by nature, not by Man.
地球上早期的火肯定是由大自然而不是人类引燃的。

例2. These signals are produced by colliding stars or nuclear reactions in outer space.
这些讯号是由外层空间的星球碰撞或者核反应所造成的。

例3. Natural light or "white" light is actually made up of many colors.
自然光或者"白光"实际上是由许多种颜色组成的。

例4. The behavior of a fluid flowing through a pipe is affected by a number of factors, including the viscosity of the fluid and the speed at which it is pumped.
流体在管道中流动的情况，受到诸如流体粘度、泵送速度等各种因素的影响。

例5. They may have been a source of part of the atmosphere of the terrestrial planets, and they are believed to have been the planetesimal-like building blocks for some of the outer planets and their satellites.
它们可能一直是地球行星的一部分大气的来源。它们还被认为是构成外部行星以及其卫星的一种类似微星的基础材料。

例6. Over the years, tools and technology themselves as a source of fundamental innovation have largely been ignored by historians and philosophers of science.
工具和技术本身作为根本性创新的源泉多年来在很大程度上被科学史学家和科学思想家们忽视了。

例7. Whether the Government should increase the financing of pure science at the expense of technology or vice versa（反之）often depends on the issue of which is seen as the driving force.
政府是以减少技术的经费投入来增加纯理论科学的经费投入，还是相反，这往往取决于把哪一方看作是驱动的力量。

例8. The supply of oil can be shut off unexpectedly at any time, and in any case, the oil wells will all run dry in thirty years or so at the present rate of use.

石油的供应可能随时会被中断；不管怎样，以目前的这种消费速度，只要 30 年左右，所有的油井都会枯竭。

Exercises

I. List three types of common operating system platform

1. _____
2. _____
3. _____

II. Describe each of the following terms

1. Operating system
2. Static resource allocation
3. Concurrent sharing

III. List the two strategies for resource allocation

IV. List two ways of sharing resource.

V. Compare three different OS platforms, mentioned in 5.2, in reliability, performance and security

Chapter 6 Software Engineering

学习指导

软件是各种各样操作计算机与有关设备的程序的总称，通常分为应用软件和系统软件两类。软件工程是硬件与系统工程发展的产物，它包括：方法、工具和过程三个要素。软件生命周期是软件工程中应用最广的范例，它要求软件开发按照系统的、顺序的方式，从系统分析开始向软件需求分析、设计、编码、测试和维护的方向发展。原型法在软件开发中被认为是通向应用的捷径。通过本章学习，读者应掌握：

- 软件的分类和基本概念等术语和知识；
- 软件的生命周期的基本术语与知识；
- 原型法的基本特点；
- 长定语从句的翻译技巧。

6.1 Basic Software Concepts

Software is a general term for the various kinds of programs used to operate computers and related devices. The term hardware describes the physical aspects of computers and related devices.

Software can be thought of as the variable part of a computer and hardware the invariable part. Software is often divided into application software (programs that do work users are directly interested in) and system software (which includes operating systems and many programs). The term middleware is sometimes used to describe programming that mediates between application and system software or between two different kinds of application software (for example, converting data from one file format to another file format).

An additional and difficult-to-classify category of software is the utility, which is a small useful program with limited capability. Some utilities come with operating systems. Like application, utilities tend to be separately installable and capable of being used independently of the rest of the operating system.

Software can be purchased or acquired as shareware (usually intended for sale after a trial period), liteware (shareware with some capability disable), freeware (free software but with copyright restrictions), public domain software (free with no restriction), and free software (software whose users agree not to limit its further distribution).

Software is usually packaged on CD-ROM and diskettes. Today, much purchased software, shareware, and freeware is downloaded over the Internet.

6.1.1 Application Software

Application software might be described as "end-user" software. Application software performs useful work on general-purpose tasks such as Word and Spreadsheet processing.

Some general kinds of application software include:
- Productivity software, which includes word processors, spreadsheets, and tools for use by most computer users.
- Presentation software, such as Microsoft PowerPoint.
- Graphics software for graphic designers, including Photoshop, 3Dmax, etc.
- CAD/CAM software.
- Specialized scientific applications.

Application software may be packaged or custom-made.
- Packaged software refers to programs prewritten by professional programmers that are typically offered for sale on a diskette. There are over 12, 000 different types of application packages available for microcomputers alone.
- Custom-made software, or custom programs, is what all software used to be. Twenty years ago organizations hired computer programmers to create all their software. The programmer custom-wrote programs to instruct the company computer to perform whatever tasks the organization wanted. A program might compute payroll checks, keep track of goods in the warehouse, calculate sales commissions, or perform similar business tasks.

There are certain general-purpose programs that we call "basic tools". These programs are widely used in nearly all career areas. They are the kind of programs you have to know to be considered computer competent.

The most popular so-called basic tool include:
- Word processing programs, used to prepare written documents; Electronic spreadsheets, used to analyze and summarize data.
- Database managers, used to organize and manage data and information.
- Graphics programs, used to visually analyze and present data and information.
- Communication programs, used to transmit and receive data and information.
- Integrated programs, which combine some or all of these applications in one program.

6.1.2 System Software

The user interacts with application software. System software enables the application software to interact with the computer. System software is "background" software, including programs that help the computer manage its own internal resources.

The most important system software program is the operating system, which interacts between the application software and the computer. The operating system handles such details as running ("executing") programs, storing data and programs, and processing ("manipulating") data. System software frees users to concentrate on solving problems rather than on the complexities of operating the computer.

Microcomputer operating systems change as the machines themselves become more powerful and outgrow the older operating systems. Today's computer competency, then, requires that you have some knowledge of the following most popular microcomputer operating systems:

- DOS, the standard operating system for personal computers manufactured by International Business Machines (IBM) and IBM-compatible microcomputers.
- Windows, which, initially, is not an operating system, but an environment that extends the capability of DOS.
- Windows NT, a powerful operating system designed for powerful microcomputers.
- Os/2, the operating system developed for IBM's more powerful microcomputers.
- Macintosh operating system, the standard operating system for Apple Corporation's Macintosh computers.
- Unix, an operating system originally developed for minicomputers. Unix is now important because it can run on many of the more powerful microcomputers.

New Words & Expressions

middleware　n. 中间（软）件
liteware　n.（light ware）不具备部分性能的共享软件
freeware　n. 免费软件
CAD/CAM　计算机辅助设计/制造
payroll　n. 工资单
Unix　n. UNIX 操作系统，一种多用户的计算机操作系统
difficult-to-classify　难以分类
shareware　n. 共享软件
spreadsheet　n. 电子制表软件，电子表格
custom-made　a. 定做的，订制的
general-purpose programs　通用程序

6.2　The Software Life Cycle

An early definition of software engineering was proposed by Fritz Bauer at the first major conference dedicated to the subject:

The establishment and use of sound engineering principles in order to obtain economically software that is reliable and works efficiently on real machines.

Although many more comprehensive definitions have been proposed, all reinforce the importance of engineering discipline in software development.

Software engineering is an outgrowth of hardware and system engineering. It encompasses a set of three key elements: methods, tools, and procedures—that enable the manager to control the process of software development and provide the practitioner with a foundation for building high-quality software in a productive manner.

Fig. 6-1 illustrates the classic life cycle paradigm for software engineering. Sometimes called the "waterfall model", the life cycle paradigm demands a systematic, sequential approach to software development that begins at the system level and progresses through analysis, design, coding, testing, and maintenance. Modeled after the conventional engineering cycle, the life cycle paradigm encompasses the following activities:

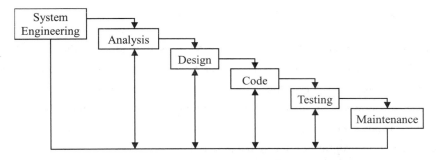

Fig. 6-1　The classic life cycle

6.2.1　System engineering and analysis

Because software is always part of a larger system, work begins by establishing requirements for all system elements and then allocating some subset of these requirements to software. This system view is essential when software must interface with other elements such as hardware, people, and databases. System engineering and analysis encompass requirements gathering at the system level with a small amount of top-level design and analysis.

6.2.2　Software requirements analysis

The requirements gathering process is intensified and focuses specifically on software. To understand the nature of the program(s)to be built, the software engineer ("analyst") must understand the information domain for the software, as well as required function, performance, and interfacing. Requirements for both the system and the software are documented and reviewed with the customer.

6.2.3　Design

Software design is actually a multistep process that focuses on three distinct attributes of the program: data structure, software architecture, and procedural detail. The design process translates requirements into a representation of the software that can he assessed for quality before coding begins. Like requirements, the design is documented and becomes part of the software configuration.

6.2.4　Coding

The design must be translated into a machine-readable form. The coding step performs this task. If design is performed in a detailed manner, coding can be accomplished mechanistically.

6.2.5　Testing

Once code has been generated, program testing begins. The testing process focuses on the logical internals of the software, assuring that all statements have been tested, and on the functional externals, that is, conducting tests to assure that defined input will produce actual results that agree with required results.

6.2.6 Maintenance

Software will undoubtedly undergo change after it is delivered to the customer (a possible exception is embedded software). Change will occur because errors have been encountered, because the software must be adapted to accommodate changes in its external environment (e.g. a change required because of a new operating system or peripheral device), or because the customer requires functional or performance enhancements. Software maintenance applies each of the preceding life cycle steps to an existing program rather than a new one.

The classic life cycle is the oldest and the most widely used paradigm for software engineering. However, over the past few years, criticism of the paradigm has caused even active supporters to question its applicability in all situations. Among the problems that are sometimes encountered when the classic life cycle paradigm is applied are:

- Real projects rarely follow the sequential flow that the model proposes. Iteration always occurs and creates problems in the application on the paradigm.
- It is often difficult in the beginning for the customer to state all requirements explicitly. The classic life cycle requires this and has difficulty accommodating the natural uncertainty that exists at the beginning of many projects.
- The customer must have patience. A working version of the program(s) will not be available until late in the project time span. A major blunder undetected until the working program is reviewed can be disastrous.

Each of these problems is real. However, the classic life cycle paradigm has a definite and important place in software engineering work. It provides a template into which methods for analysis, design, coding, testing, and maintenance can be placed. The classic life cycle remains the most widely used procedural model for software engineering. While it does have weaknesses, it is significantly better than a haphazard approach to software development.

New Words & Expressions

software life cycle 软件生存期
comprehensive adj. 全面的
discipline n. 学科
paradigm n. 范例
sequential adj. 顺序的
software requirement analysis 软件需求分析
iteration n. 迭代, 反复

software engineering 软件工程
reinforce n. 加强, 强调
outgrowth n. 派生物
waterfall model 瀑布方式
system engineering and analysis 系统工程与分析
maintenance n. 维护
template n. 模板

Notes

1. The establishment and use of sound engineering principles in order to obtain economically software that

is reliable and works efficiently on real machines. 这是一个组合名词，而不是一个句子，意为：建立和使用稳妥的工程原理以期得到一种能在实际机器上可靠而高效工作的、经济的软件。其中，in order to 是状语，修饰 principle，that 引导一个定语从句修饰 software。

2. The requirements gathering process is intensified and focuses specifically on software. 需求分析的收集过程特别重要，并且着重从软件的角度进行收集。

6.3 Prototyping

Prototyping attacks the design problem by giving system analysts and users opportunities to catch design glitches before they are set in hard code.

A prototype usually begins taking shape as a system analyst and a user work out a plan on paper. Next, using a development tool, the analyst maps out the application. Finally, the user and the analyst work with the screens, moving or changing elements or highlighting them to make the application more workable.

Prototypes are usually constructed with tools are designed for broader tasks, such as application generators. Other software packages commonly called programmer productivity tools, are also appropriate for the task, including fourth-generation languages, documentation generators, and system-development tools.

Prototyping can directly cut the cost of producing applications by eliminating the fix-it coding common to traditionally developed system. In addition, it can dramatically raise the productivity of data center staffers, indirectly saving money.

Some proponents believe failed prototypes afford users an educational experience. Users in their construction learn first hand about the problem that MIS/DP has been trying to explain. Many prototypes end up in the trash bin, but experimentation is part of the game and shouldn't be considered a drawback. Organizations must be willing to sacrifice a certain amount of time and effort to attain the benefits of prototyping, and those benefits outweigh the false starts.

When making prototyping part of the development process, don't try to prototype systems that tie together transactions in several departments; start with something simple and self-contained that doesn't have to be integrated into another system such as a marketing-information system.

Enlist user members of project team carefully. The higher the rank of the user members, the better the system will be. Not all programmers and analysts are suited to be prototypers. They must work closely with users, a role that some technicians find difficult.

Fostering better relations with users won't be directly translated into an improvement of the bottom line. Nevertheless, prototypes may help to improve a corporation's ability to compete by reducing the amount of time needed to generate vital new information systems.

New Words and Expressions

prototyping   n. 原型法
set in   v. 开始，到来，插入

glitch   n. 小故障，小毛病
prototype   n. 原型

take shape	v. 成形，形成	work out	v. 可以解决，设计出，计算出
map out	v. 制订	highlight	v. 加亮，使显著，突出
appropriate for	适于，合乎	fix-it coding	纠错式的编码，修改式的编码
staffers	n. 编辑，职员	end up	v. 结束
outweigh	v. 重量（或价值等）上超过		

Abbreviations

MIS/DP （Management Information System/Data Processing）管理信息系统/数据处理

Notes

1. Prototyping attacks the design problem by giving system analysts and users opportunities to catch design glitches before they are set in hard code. 原型法这样处理设计问题：在开始艰难的编码之前，为系统分析员和用户提供一个机会来捕捉设计中的毛病。
2. A prototype usually begins taking shape as a system analyst and a user work out a plan on paper. 当系统分析员和用户设计出一个计划时，首先构造一个原型。

Reading Material: Software Engineering Methodologies

Early approaches to software engineering insisted on performing requirements analysis, design, implementation, and testing in a strictly sequential manner. The belief was that too much was at risk during the development of a large software system to allow for variations. As a result, software engineers insisted that the entire requirements specification of the system be completed before beginning the design and, likewise, that the design be completed before beginning implementation. The result was a development process now referred to as the **waterfall model,** an analogy to the fact that the development process was allowed to flow in only one direction.

In recent years, software engineering techniques have changed to reflect the contradiction between the highly structured environment dictated by the waterfall model and the "free-wheeling," trial-and-error process that is often vital to creative problem solving. This is illustrated by the emergence of the **incremental model** for software development. Following this model, the desired software system is constructed in increments—the first being a simplified version of the final product with limited functionality. Once this version has been tested and perhaps evaluated by the future user, more features are added and tested in an incremental manner until the system is complete. For example, if the system being developed is a patient records system for a hospital, the first increment may incorporate only the ability to view patient records from a small sample of the entire record system. Once that version is operational, additional features, such as the ability to add and update records, would be added in a stepwise manner.

Another model that represents the shift away from strict adherence to the waterfall model is the **iterative model,** which is similar to, and in fact sometimes equated with, the incremental model, although the two are distinct. Whereas the incremental model carries the notion of *extending* each preliminary version of a product into a larger version, the iterative model encompasses the concept of *refining* each version. In reality, the incremental model involves an underlying iterative process, and the iterative model may incrementally add features.

A significant example of iterative techniques is the **rational unified process** (**RUP,** rhymes with "cup") that was created by the Rational Software Corporation, which is now a division of IBM. RUP is essentially a software development paradigm that redefines the steps in the development phase of the software life cycle and provides guidelines for performing those steps. These guidelines, along with CASE tools to support them, are marketed by IBM. Today, RUP is widely applied throughout the software industry. In fact, its popularity has led to the development of a nonproprietary version, called the **unified process,** that is available on a noncommercial basis.

Incremental and iterative models sometimes make use of the trend in software development toward **prototyping** in which incomplete versions of the proposed system, called prototypes, are built and evaluated. In the case of the incremental model these prototypes evolve into the complete, final system—a process known as **evolutionary prototyping.** In a more iterative situation, the prototypes may be discarded in favor of a fresh implementation of the final design. This approach is known as **throwaway prototyping.** An example that normally falls within this throwaway category is **rapid prototyping** in which a simple example of the proposed system is quickly constructed in the early stages of development. Such a prototype may consist of only a few screen images that give an indication of how the system will interact with its users and what capabilities it will have. The goal is not to produce a working version of the product but to obtain a demonstration tool that can be used to clarify communication between the parties involved in the software development process. For example, rapid prototypes have proved advantageous in clarifying system requirements during requirements analysis or as aids during sales presentations to potential clients.

A less formal incarnation of incremental and iterative ideas that has been used for years by computer enthusiasts/hobbyists is known as **open-source development.** This is the means by which much of today's free software is produced. Perhaps the most prominent example is the Linux operating system whose open-source development was originally led by Linus Torvalds. The open-source development of a software package proceeds as follows: A single author writes an initial version of the software (usually to fulfill his or her own needs) and posts the source code and its documentation on the Internet. From there it can be downloaded and used by others without charge. Because these other users have the source code and documentation, they are able to modify or enhance the software to fit their own needs or to correct errors that they find. They report these changes to the original author, who incorporates them into the posted version of the software, making this extended version available for further modifications. In practice, it is possible for a software package to evolve through several extensions in a single week.

Perhaps the most pronounced shift from the waterfall model is represented by the collection of methodologies known as **agile methods,** each of which proposes early and quick implementation on an incremental basis, responsiveness to changing requirements, and a reduced emphasis on rigorous requirements analysis and design. One example of an agile method is **extreme programming (XP).**

Following the XP model, software is developed by a team of less than a dozen individuals working in a communal work space where they freely share ideas and assist each other in the development project. The software is developed incrementally by means of repeated daily cycles of informal requirements analysis, designing, implementing, and testing. Thus, new expanded versions of the software package appear on a regular basis, each of which can be evaluated by the project's stakeholders and used to point toward further increments. In summary, agile methods are characterized by flexibility, which is in stark contrast to the waterfall model that conjures the image of managers and programmers working in individual offices while rigidly performing well-defined portions of the overall software development task.

The contrasts depicted by comparing the waterfall model and XP reveal the breadth of methodologies that are being applied to the software development process in the hopes of finding better ways to construct reliable software in an efficient manner. Research in the field is an ongoing process. Progress is being made, but much work remains to be done.

New Words & Expressions

Methodology n. 方法学	insist vi. 坚持、强调
free-wheeling 随心所欲的	trial-and-error 实验和误差
incremental adj. 增加的	adherence n. 遵循，粘附
iterative adj. 反复的，迭代的；n. 反复体	rhyme vi. 押韵，和…同韵
rational adj. 理性的；理智的；合理的	unified adj. 统一的，统一标准的
prototyping n. 原型机制造	throwaway n. 广告传单，散单，宣传小册子
incarnation n. 体现，具体化，典型，化身	agile adj. 敏捷的，灵活的
extreme adj. 末端的，极端的	

长定语（从句）的翻译技巧之一

科技英语中，句子的某个中心词常常被若干个后置定语或定语从句所修饰。这是因为英语句法结构重"形合"，任何一个名词或名词词组，为了对它进行完整而明确的阐述，可以借助各种修饰结构（介词短语、不定式短语、分词短语、定语或同位语从句等）像滚雪球那样引出一长串修饰成分。而汉语讲"意合"，各个成分用意义串联，往往不需要连接词，一般句子较短。汉语中，对于一个名词，通常不用"叠床架屋"式的多重修饰语，因为过长的前置定语会使句子拖泥带水、冗长臃肿、晦涩难懂。

由于英汉两种语言的修饰方式有很大差别，所以对于科技英语中含有多重定语或长定语从句的中心词（组）没有现成的对应译法，也就难以用规范的汉语，简练地译出原文复杂的修饰关系。

实际上，翻译复杂的多重定语（从句），不但许多初学英语者视为畏途，常常感到"只能意会，难以言传"，就是对于具有一定经验的译者，也是一个难题。例如：

例1. A stack is a data type whose major attributes are determined by the rules governing the insertion and deletion of its elements.

译文：栈是主要性质由支配其元素的插入与删除的规则来决定的一种数据类型。

译文中，动宾之间所含内容过多，形成"大肚子"句，行文不畅。

改译：栈是这样一种数据类型，其主要性质由支配其元素的插入与删除的规则来决定。

例 2. A fast and accurate symbol manipulating system that is organized to accept, store, and process data and produce output results under the direction of a stored program of instruction is a computer.

原译：一种能够接收、存储、和处理数据，并能在存储指令程序控制下产生输出结果的快速而准确地处理符号的系统叫做计算机。

由于译文主语较长，所以句子显得有点"头重脚轻"，读起来感到不够顺畅。

改译：一种快速而准确地处理符号的系统叫做计算机，它能够接收、存储、和处理数据，并能在存储指令程序控制下产生输出结果。

由上述一些例句可以发现，翻译科技英语中含有复杂定语或长定语从句的中心词（组）时一般容易犯的"通病"是：把原文机械地译成汉语的偏正词组，以致出现"重负荷"修饰语，影响译文的畅达。下文讲述一些技巧，在不少情况下，可以避免译文中出现"头重脚轻"的现象。

一、先提后述法

此译法适用于结构复杂的并列长定语（从句）。翻译时，首先用"这样（的）""这样一些""下列（的）"等词语概括所有修饰成分，然后分别叙述各个修饰内容。用这种方法翻译的译文，重点突出、条理清楚。在许多情况下，这是一种翻译并列长定语（从句）行之有效的方法。例如：

例 1. A computer is an electronic device that can receive a set of instructions, or program, and then carry out this program by performing calculations on numerical data or by compiling and correlating other forms of information.

译文：计算机是一种电子装置，它能接受一套指令或程序，并通过数据运算，或收集和联系其他形式的信息来执行该程序。

（比较：计算机是一种能接受一套指令或程序，并通过数据运算，或收集和联系其他形式的信息来执行该程序的电子装置）

例 2. A sound card is a printed circuit board that can translate digital information into sound and back, that plug into a slot on the motherboard (the main circuit board of a computer) and is usually connected to a pair of speakers.

译文：声卡是一块印刷电路板，它能把数字信息译为声音，也能把声音变为数字信息，它插在母板（计算机主电路板）上的槽内，而且通常连接一对喇叭。

（比较：声卡是一块能把数字信息译为声音，也能把声音变为数字信息，插在母板（计算机主电路板）上的槽内，而且通常连接一对喇叭的印刷电路板。）

例 3. The transistor meant more powerful, more reliable, and less expensive computers that would occupy less space and give off less heat than did vacuum-tube-powered computers.

译文：使用晶体管可做成功能更强、更可靠、更价廉的计算机，它与真空管计算机相比，占地面积小，产生的热量少。

（比较：使用晶体管可做成功能更强、更可靠、更价廉的，与真空管计算机相比，占地面积小，产生的热量小的计算机。）

例4. The World Wide Web is one of the Internet's most popular services, providing access to over one billion Web pages, which are documents created in a programming language called HTML and which can contain text, graphics, audio, video, and other objects, as well as "hyperlinks" that permit a user to jump easily from one page to another.

万维网是因特网最流行的服务之一，提供对超过10亿网页的访问，这些网页是由一种叫做HTML（超文本链接标示语言）的编程语言生成的文件，它们可以包含本文、图形、声频、视频和其他的对象、以及允许用户容易地跳跃到其他网页的"超链接"。

二、先述后提法

此译法是"先提后述法"的倒置。也就是先叙述中心词（组）的修饰内容，最后用"这样的""这一切""这种""这些"等词语予以呼应。对于科技英语文献，这往往也是一种翻译长定语或定语从句的有效方法。例如：

例1. The fact that the Government may have formulated, furnished, or in any way supplied the said drawings, specifications, or other data is not to be regarded by implication ⋯

译文：政府可能已经系统地阐述、提供或以某种方式供应了上述图纸、规范或其他资料，但这样的事实并非暗示……

（比较：政府可能已经系统地阐述、提供或以某种方式供应了上述图纸、规范或其他资料的这一事实，并非暗示是……）

例2. The way in which a number of processor units are employed in a single computer system to increase the performance of the system in its application environment above the performance of single processor is an organizational technique.

译文：将许多处理器组成单一的计算机系统，从而提高该系统在其运行环境中的性能，使其超过单个处理器的性能，这种方法称为组织技术。

（比较：将许多处理器组成单一的计算机系统，从而提高该系统在其运行环境中的性能，使其超过单个处理器的性能的方法称为组织技术。）

例3. The developer must be aware of certain basic acoustic principles which must all be taken into account during the design stages of the office and which, at certain points, can conflict other requirements posed by the office layout, decor, environmental control, etc.

译文：某些基本的声学因素，在办公室的设计阶段就必须予以考虑，因为它们在某些方面可能与办公室的布局、室内装饰、环境控制等方面的其他要求相抵触。因此，对于这些声学因素，设计人员务必注意。

（比较：设计人员必须注意在设计阶段应该予以考虑的并且在某些方面可能与办公室的布局、室内装饰、环境控制等方面的其他要求相抵触的某些基本声学因素。）

例4. Each publication maintains its own editorial policy but, by and large, articles that report research results from a practical viewpoint or describe application of new technology or procedures to existing problem merit publication.

译文：每种出版物都有各自的编辑原则，但一般来说，从实用观点报道研究成果，或描述新技术的应用，或描述存在问题的过程，这样的文章都是值得出版的。

（比较：每种出版物都有各自的编辑原则，但一般来说，文章从实用观点报道研究成果，或描述新技术的应用，或描述存在问题的过程是值得出版的。）

Exercises

I. Fill in the following blanks

1. The _____ system is the physical equipment that you can see and touch .
2. Software is often divided into _____ software and _____ software .
3. _____ software controls the computer and enables it to run the _____ software and hardware .
4. _____ means building a small version of a system, usually with limited functionality that can be used to help the user or customer identify the key requirements of a system and demonstrate feasibility of a design or approach .

II. Describe the stages involved in the software life cycle

III. Answer the following Questions

1. Summarize the distinction between the traditional waterfall model of software development and the newer incremental and iterative paradigms.
2. Identify three development paradigms that represent the move away from strict adherence to the waterfall model.

Chapter 7 Programming Language

本章学习指导

程序设计语言是指令计算机实现某些具体任务的一套特殊词汇和一组语法规则。通常，程序设计语言是指高级语言，在高级语言和机器语言之间的是汇编语言。面向对象的程序设计是一种以面向"对象"的，而不是"行为"的，面向数据而不是逻辑的方式组织的程序设计语言模式。通过本章的学习，读者应掌握：
- 机器语言、汇编语言和高级语言的特点；
- 面向对象的程序设计（OOP）的概念和特征；
- 统一建模语言（UML）的特点；
- 复杂从句的翻译技巧。

7.1 Introduction to Programming Language

A programming language represents a special vocabulary and a set of grammatical rules for instructing a computer to perform specific tasks. Broadly speaking, it consists of a set of statements or expressions understandable to both people and computers. People understand these instructions because they use human (English and mathematical) expressions. Computers, on the other hand, process these instructions through use of special programs, which, known as translators, decode the instructions from people and create machine-language coding.

The term programming language usually refers to high-level languages, such as BASIC, C, C++, COBOL, FORTRAN, Ada and Pascal. Each language has a unique set of key words (words that it understands) and a special syntax for organizing program instructions.

High-level programming languages, while simple compared to human languages, are more complex than the languages the computer actually understands, which are called machine languages. Each different type of CPU (Central Processing Unit) has its own unique machine language.

Lying between machine languages and high-level languages are assembly languages, which are directly related to a computer's machine language. In other words, it takes one assembly command to generate each machine-language command. Machine languages consist entirely of numbers and are almost impossible for humans to read and write. Assembly languages have the same structure and set of commands as machine languages, but they enable a programmer to use names instead of numbers.

Each type of CPU has its own machine language and assembly language, so an assembly language program written for one type of CPU won't run on another. In the early days of programming, all programs were written in assembly languages. Now, most programs are written in a high-level language such as FORTRAN or C. Programmers still use assembly languages when speed is essential or when they need to perform an operation that isn't possible in a high-level language.

Lying above high-level languages are those called fourth-generation languages (usually abbreviated 4GL). 4GLs are far removed from machine languages and represent the class of computer languages closest to human languages. Most 4GLs are used to access databases. For example, a typical 4GL command is: FIND ALL RECORDS WHERE NAME IS "SMITH".

More recently, a new type of programming language has emerged that supports Object Oriented Programming (OOP), including C++, Microsoft Visual C++, Visual Foxpro, and Visual Java. OOP is a type of programming in which programmers define not only the data type of a data structure, but also the types of operations (functions, or methods) that can be applied to the data structure. In this way, the data structure becomes an object that includes both data and functions. In addition, programmers can create relationships between one object and another. For example, objects can inherit characteristics from other objects. One of the principal advantages of object-oriented programming techniques over procedural programming techniques is that they enable programmers to create modules that do not need to be changed when a new type of object is added. A programmer can simply create a new object that inherits many of its features from existing objects. This makes object-oriented programs easier to modify. To perform object-oriented programming, one needs an object-oriented programming language (OOPL). C++ and Small talk are two of the more popular languages, and there are also object-oriented versions of Pascal.

All high-level language programs must be translated into machine language so that the computer can understand it. There are two ways to do this: compile the program or interpret the program.

The question of which language is best is one that consumes a lot of time and energy among computer professionals. Every language has its strengths and weaknesses. For example, FORTRAN is a particularly good language for processing numerical data, but it does not lend itself very well to organizing large programs. Pascal is very good for writing well-structured and readable programs, but it is not as flexible as the C programming language. C++ embodies powerful object-oriented features, but it is complex and difficult to learn. The choice of which language to use depends on the type of computer the program is to run on, what sort program it is, and the expertise of the programmer.

The trend toward higher-level languages, initiated in the 1950s, is still continuing. Today, there is talk about natural language communication between people and computers. That is, tools are being developed to make it possible for people to address computers in normal, conversational language. Both written and spoken version of natural languages is in use. However, a great deal of developmental effort remains before these languages become common in business applications.

New Words & Expressions

translator n. 翻译器，翻译程序   syntax n. 语法，句法

machine language 机器语言　　　assembly language 汇编语言
abbreviate v. 缩短，缩写　　　compile v. 编译
interpret v. 解释

Abbreviations

OOPL (object-oriented programming language) 面向对象的程序设计语言

Notes

1. OOP is a type of programming in which programmers define not only the data type of a data structure, but also the types of operations that can be applied to the data structure. OOP 这种程序设计要求编程人员不仅要给出数据结构中的数据类型的定义，还需要给出作用在这些数据结构之上的操作（函数，或方法）的类型。

7.2　Object-oriented Programming

Object-oriented programming (OOP) is a programming language model organized around "objects" rather than "actions" and data rather than logic. Historically, a program has been viewed as a logical procedure that takes input data, processes it, and produces output data. The programming challenge was seen as how to write the logic, not how to define the data. Object-oriented programming takes the view that what we really care about are the objects we want to manipulate rather than the logic required to manipulate them. Examples of objects range from human beings (described by name, address, and so forth) to buildings and floors (whose properties can be described and managed) down to the little widgets on your computer desktop (such as buttons and scroll bars).

The first step in OOP is to identify all the objects you want to manipulate and how they relate to each other, an exercise often known as data modeling. Once you've identified an object, you generalize it as a class of objects (think of Plato's concept of the "ideal" chair that stands for all chairs) and define the kind of data it contains and any logic sequences that can manipulate it. Each distinct logic sequence is known as a method. A real instance of a class is called (no surprise here) an "object" or, in some environments, an "instance of a class." The object or class instance is what you run in the computer. Its methods provide computer instructions and the class object characteristics provide relevant data. You communicate with objects - and they communicate with each other - with well-defined interfaces called *messages*.

The concepts and rules used in object-oriented programming provide these important benefits:
- The concept of a data class makes it possible to define subclasses of data objects that share some or all of the main class characteristics. Called inheritance, this property of OOP forces a more thorough data analysis, reduces development time, and ensures more accurate coding.
- Since a class defines only the data it needs to be concerned with, when an instance of that

class (an object) is run, the code will not be able to accidentally access other program data. This characteristic of data hiding provides greater system security and avoids unintended data corruption.
- The definition of a class is reuseable not only by the program for which it is initially created but also by other object-oriented programs (and, for this reason, can be more easily distributed for use in networks).
- The concept of data classes allows a programmer to create any new data type that is not already defined in the language itself.

One of the first object-oriented computer languages was called Smalltalk. C++ and Java are the most popular object-oriented languages today. The Java programming language is designed especially for use in distributed applications on corporate networks and the Internet.

New Words & Expressions

widget   n. 小器具，小部件
object   n. 对象
method   n. 方法
inheritance   n. 继承，继承性
data modeling   数据建模
class   n. 类
relevant   adj. 相关的
data hiding   数据隐藏

7.3 OMG's Unified Modeling Language(UML)

Large enterprise applications—the ones that execute core business applications, and keep a company going—must be more than just a bunch of code modules. They must be structured in a way that enables scalability, security, and robust execution under stressful conditions, and their structure—frequently referred to as their *architecture*—must be defined clearly enough that maintenance programmers can (quickly!) find and fix a bug that shows up long after the original authors have moved on to other projects. That is, these programs must be *designed* to work perfectly in many areas, and business functionality is not the only one (although it certainly is the essential core). Of course a well-designed architecture benefits any program, and not just the largest ones as we've singled out here. We mentioned large applications first because structure is a way of dealing with complexity, so the benefits of structure (and of modeling and design, as we'll demonstrate) compound as application size grows large. Another benefit of structure is that it enables *code reuse*: Design time is the easiest time to structure an application as a collection of self-contained modules or components. Eventually, enterprises build up a library of models of components, each one representing an implementation stored in a library of code modules. When another application needs the same functionality, the designer can quickly import its module from the library. At coding time, the developer can just as quickly import the code module into the executable.

Modeling is the designing of software applications before coding. Modeling is an Essential Part of large software projects, and helpful to medium and even small projects as well. A model plays the analogous role in software development that blueprints and other plans (site maps, elevations, physical models) play in the building of a skyscraper. Using a model, those responsible for a software development project's success can assure themselves that business functionality is complete and correct, end-user needs are met, and program design supports requirements for scalability, robustness, security, extendibility, and other characteristics, before implementation in code renders changes difficult and expensive to make. Surveys show that large software projects have a huge probability of failure - in fact, it's more likely that a large software application will fail to meet all of its requirements on time and on budget than that it will succeed. If you're running one of these projects, you need to do all you can to increase the odds for success, and modeling is the only way to visualize your design and check it against requirements before your crew starts to code.

***The OMG's Unified Modeling Language*™(UML®)**helps you specify, visualize, and document models of software systems, including their structure and design, in a way that meets all of these requirements. (You can use UML for business modeling and modeling of other non-software systems too.) Using any one of the large number of UML-based tools on the market, you can analyze your future application's requirements and design a solution that meets them, representing the results using UML's twelve standard diagram types.

You can model just about any type of application, running on any type and combination of hardware, operating system, programming language, and network, in UML. Its flexibility lets you model distributed applications that use just about any middleware on the market. Built upon the MOF™ metamodel which defines class and operation as fundamental concepts, it's a natural fit for object-oriented languages and environments such as C++, Java, and the recent C#, but you can use it to model non-OO applications as well in, for example, Fortran, VB, or COBOL.UML Profiles (that is, subsets of UML tailored for specific purposes) help you model Transactional, Real-time, and Fault-Tolerant systems in a natural way.

You can do other useful things with UML too: For example, some tools analyze existing source code (or, some claim, object code!) and reverse-engineer it into a set of UML diagrams. Another example: In spite of UML's focus on design rather than execution, some tools on the market *execute* UML models, typically in one of two ways: Some tools execute your model interpretively in a way that lets you confirm that it really does what you want, but without the scalability and speed that you'll need in your deployed application. Other tools (typically designed to work only within a restricted application domain such as telecommunications or finance) generate program language code from UML, producing most of a bug-free, deployable application that runs quickly if the code generator incorporates best-practice scalable patterns for, e.g., transactional database operations or other common program tasks. Our final entry in this category: A number of tools on the market generate Test and Verification Suites from UML models.

***UML and OMG's Model Driven Architecture*™ (MDA™):** A few years ago (in fact, surprisingly few!), the biggest problem a developer faced when starting a distributed programming project was finding a middleware with the functionality that he needed, that ran on the hardware and operating systems running in his shop. Today, faced with an embarrassingly rich array of middleware platforms, the developer has three different middleware problems: First, selecting one; second, getting it to work with the other platforms already deployed not only in his own shop, but also those of his customers and suppliers; and third, interfacing to (or, worse yet, migrating to) a new "Next Best Thing" when a new platform comes along and catches the fancy of the analysts and, necessarily, CIOs everywhere.

New Words & Expressions

Scalability   n. 可扩展性
Bug   n. 故障，错误
import   v. 引入，导入，进口
end-user   n.[计] 终端用户
odd   n. 出乎意外的事
middleware   n. 中间件
transactional   adj. 事务性的
fault-tolerant   adj. 容错的
incorporate   v. 结合，合并

robust   adj. 健壮的
self-contained   adj. 自主的，独立的
analogous   adj. 类似的，相似的
robustness   n. 鲁棒性，健壮性
flexibility   n. 灵活性
metamodel   n. 原型
real-time   adj. 实时的
reverse-engineer   v. 反编译
deploy   v. 部署

Abbreviations

UML (Unified Modeling Language)   统一建模语言
MDA (Model Driven Architecture)   模型驱动的结构
non-OO (non-Object Oriented)   非面向对象的

Reading Material: Programming Paradigms

The generation approach to classifying programming languages is based on a linearscale on which a language's position is determined by the degreeto which the user of the language is freed from the world of computer gibberishand allowed to think in terms associated with the problem being solved. In reality,the development of programming languages has not progressed in this manner but has developed along different paths as alternative approaches to the programmingprocess (called **programming paradigms**) have surfaced and beenpursued. Consequently, the historical development of programming languages isbetter represented by a multiple-track diagram as shown in Fig.7-1, in which different paths resulting from different paradigms are shown to emerge andprogress independently. In particular, the figure presents four paths representingthe functional, object-oriented, imperative, and declarative paradigms, with various languages associated with each paradigm positioned in a manner that indicates their births relative to other languages. (It does not imply that one language necessarily evolved from a previous one.)

We should note that although the paradigms identified in Fig.7-1arecalled *programming* paradigms, these alternatives have ramifications beyond theprogramming process. They represent fundamentally different approaches tobuilding solutions to problems and therefore affect the entire software development process. In this sense, the term *programming paradigm* is a misnomer. Amore realistic term would be *software development paradigm*.

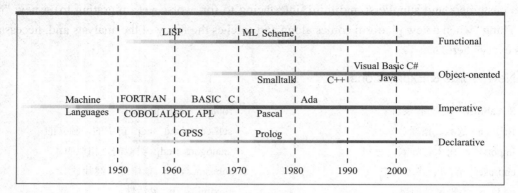

Fig.7-1　The evolution of programming paradigms

The **imperative paradigm,** also known as the **procedural paradigm,** represents the traditional approach to the programming process. It is the paradigmon which our pseudocode as well as the machine language is based. As the name suggests, the imperative paradigm defines the programming process to be the development of a sequence of commands that, when followed, manipulate data to produce the desired result. Thus the imperative paradigm tells us to approach the programming process by finding analgorithm to solve the problem at hand and then expressing that algorithm as asequence of commands.

In contrast to the imperative paradigm is the **declarative paradigm,** which asks a programmer to describe the problem to be solved rather than an algorithm to be followed. More precisely, a declarative programming system applies a preestablished general-purpose problem-solving algorithm to solve problem presented to it. In such an environment the task of a programmer becomes that of developing a precise statement of the problem rather than of describing analgorithm for solving the problem.

A major obstacle in developing programming systems based on the declarative paradigm is the need for an underlying problem-solving algorithm. For thisreason early declarative programming languages tended to be special-purpose innature, designed for use in particular applications. For example, the declarative approach has been used for many years to simulate a system (political, economic,environmental, and so on) in order to test hypotheses or to obtain predictions.In these settings, the underlying algorithm is essentially the process of simulatingthe passage of time by repeatedly recomputing values of parameters (grossdomestic product, trade deficit, and so on) based on the previously computed values.Thus, implementing a declarative language for such simulations requires that one first implement an algorithm that performs this repetitive procedure.Then the only task required of a programmer using the system is to describe the situation to be simulated. In this manner, a weather forecaster does not need to develop an algorithm for forecasting the weather but merely describes the current weather status, allowing the underlying simulation algorithm to produce weather predictions for the near future.

A tremendous boost was given to the declarative paradigm with the discovery that the subject of formal logic within mathematics provides a simple problem solving algorithm suitable for use in a general-purpose declarative programming system. The result has been increased attention to the declarative paradigm and the emergence of **logic programming**.

Another programming paradigm is the **functional paradigm.** Under this paradigm a program is viewed as an entity that accepts inputs and produces outputs. Mathematicians refer to such entities as functions, which is the reason this approach is called the functional paradigm. Under this paradigm a programis constructed by connecting smaller predefined program units (predefined functions) so that each unit's outputs are used as another unit's inputs in such away that the desired overall input-to-output relationship is obtained. In short, the programming process under the functional paradigm is that of building functions as nested complexes of simpler functions.

Note that this imperative program consists of multiple statements, each of which requests that a computation be performed and that the result be stored for lateruse. In contrast, the functional program consists of a single statement in which the result of each computation is immediately channeled into the next. In a sense, the imperative program is analogous to a collection of factories, each conver ting its raw materials into products that are stored in warehouses. From these warehouses, the products are later shipped to other factories as they are needed. But the functional program is analogous to a collection of factories that are coordinated so that each produces only those products that are ordered by other factories and then immediately ships those products to their destinations without intermediate storage. This efficiency is one of the benefits proclaimed by proponents of the functional paradigm.

Still another programming paradigm (and the most prominent one in today's software development) is the **object-oriented paradigm,** which is associated with the programming process called **object-oriented programming (OOP).**Following this paradigm, a software system is viewed as a collection of units,called **objects,** each of which is capable of performing the actions that are immediately related to itself as well as requesting actions of other objects. Together,these objects interact to solve the problem at hand.

As an example of the object-oriented approach at work, consider the task of developing a graphical user interface. In an object-oriented environment, the icons that appear on the screen would be implemented as objects. Each of these objects would encompass a collection of procedures (called **methods** in theobject-oriented vernacular) describing how that object is to respond to the occurrence of various events, such as being selected by a click of the mouse button orbeing dragged across the screen by the mouse. Thus the entire system would becon structed as a collection of objects, each of which knows how to respond to the events related to it.

To contrast the object-oriented paradigm with the imperative paradigm,consider a program involving a list of names. In the traditional imperative paradigm, this list would be merely a collection of data. Any program unitaccessing the list would have to contain the algorithms for performing there quired manipulations. In the object-oriented approach, however, the list would be constructed as an object that consisted of the list together with a collection of methods for manipulating the list. (This might include proced ures for inserting a new entry in the list, deleting an entry from the list, detecting if the list is empty, and sorting the list.) In turn, another program unit that needed to manipulate the list would not contain algorithms for performing the pertinent tasks. Instead, it would make use of the procedures provided in the object. In a sense, rather than sorting the list as in the imperative paradigm, the program unit would ask the list to sort itself.

Although we have discussed the object-oriented paradigm in more detail in Section 7.2, its significance in today's software development arena dictates that we include the concept of a class in this introduction. To this end, recall that an object can consist of data (such as a list of names) together with a collection of methods for performing activities (such as inserting new names in the list). These features must be described by statements in the written program. This description of the object's properties is called a **class.** Once a class has been constructed, it can be applied anytime an object with those characteristics is needed. Thus, several objects can be based on (that is, built from) the same class. Just like identical twins, these objects would be distinct entities but would have the same characteristics because they are constructed from the same template (the same class). (An object that is based on a particular class is said to be an **instance** of that class.)

It is because objects are well-defined units whose descriptions are isolated in reusable classes that the object-oriented paradigm has gained popularity. Indeed, proponents of object-oriented programming argue that the object-oriented paradigm provides a natural environment for the "building block" approach to software development. They envision software libraries of predefined classes from which new software systems can be constructed in the same way that many traditional products are constructed from off-the-shelf components. Building and expanding such libraries is an ongoing process.

In closing, we should note that the methods within an object are essentially small imperative program units. This means that most programming languages based on the object-oriented paradigm contain many of the features found in imperative languages. For instance, the popular object-oriented language C++ was developed by adding object-oriented features to the imperative language known as C. Moreover, since Java and C# are derivatives of C++, they too have inherited this imperative core.

New Words & Expressions

paradigm　　n. 范例；样式；模范
ramifications　n.（众多复杂而又难以预料的）结果
imperative　　adj. 必要的，紧急的，极重要的
obstacle　　n. 障碍(物)，妨碍
vernacular　　n. 方言，行话

gibberish　　n. 令人费解的话，莫名其妙的话，胡扯
misnomer　　n. 使用不当的名字或名称
procedural　adj. 程序上的
deficit　　n. 缺款额，赤字，亏损，逆差
derivative　n. 派生物，引出物

长定语（从句）的翻译技巧之二

三、解环法

这种方法适用于翻译"连环式"后置定语或定语从句，也就是当原文中的一个中心词（组）被若干个定语（从句）一环扣一环地修饰时，可先将中心词（组）译出，或把中心词（组）与靠近它的一个或两个后置定语（从句）译成汉语偏正词组，然后顺着"修饰环"依次翻译其余成分。

采用这种方法译出的句子层次分明,脉络清楚。由于科技英语中"连环式"修饰语出现较多,所以"解环法"具有较大的实用价值。例如:

例1. A flight simulator is a perfect example of programs that create a virtual reality, or computer-generated "reality" in which the user dose not merely watch but is able to actually participate.

译文:飞行模拟器是创造虚拟现实的程序的一个完美例子,或者也可以叫它计算机生成的"现实",在这个"现实"中,用户不仅能看,而且能实际参与。

(比较:飞行模拟器是创造虚拟现实,或者也可以叫它计算机生成的,用户不仅能看,而且能实际参与的"现实"的程序的一个完美例子。)

例2. In fact, the Internet is a giant example of client/server computing in which over 70 million host server computers store Web pages and other content that can be easily accessed by nearly a million local area networks and hundreds of millions of client machines worldwide.

事实上,因特网是客户机/服务器计算技术的一个巨大实例,其中,超过7000万部主机服务器计算机储存网页和其他内容,这些网页和内容能被全世界接近一百万个局域网和数亿台客户机容易地访问。

(比较:事实上,因特网是客户机/服务器计算技术的一个巨大实例,其中,超过7000万部主机服务器计算机储存能被全世界接近一百万个局域网和数亿台客户机容易地访问的网页和其他内容。)

例3. Object-oriented databases store and manipulate more complex data structures, called "object", which are organized into hierarchical classes that may inherit properties from classes higher in the chain; this database structure is the most flexible and adaptable.

译文:面向对象的数据库可以存储并处理更加复杂的数据结构,这种数据结构称为"对象","对象"可以按层次组成"类",低层的"类"可以继承上层"类"的属性;这是一种最灵活,最具适应性的数据结构。

(比较:面向对象的数据库可以存储并处理更加复杂的、称为"对象"的,可以按层次组成"类"的,低层的"类"可以继承上层"类"属性的数据结构;这是一种最灵活,最具适应性的数据结构。)

在英语科技文献中,对于某些连环式修饰语,有时还需同时借助"先提后叙法",才能使语句顺畅。例如:

例4. It (chapter 2) provides numerous solutions <u>for protecting problems associated with any type of instruments using microprocessors that must be protected when voltage supply fluctuates</u>.

就修饰关系而言,这个句子属于连环式修饰。由于各个修饰环扣得很紧——原文中,protecting problems 并非仅仅与 associated with … instruments 有关,而且与修饰 instruments 的分词短语 using microprocessors 及修饰 microprocessors 的定语从句 that … fluctuates 密切相关。(定语从句本身带有一个较长的状语)。因此,译成汉语时,为了使修饰成分紧扣修饰对象,既不宜直接译成偏正词组——这会造成"大肚子"句,见译文(1);也不宜用"解环法"法逐个翻译修饰环——这会导致修饰语与中心词关系松弛,见译文(2)。此时可采用"解环法+先提后叙法",也就是先译出一两个修饰环,再用"这样的""这样一些"等词语"扣"住中心词,最后叙述具体修饰内容。这样,不但句子修饰关系明确,而且行文较为流畅,见译文(3)。

（1）第二章提供与使用当供电电压波动时必须加以保护的微处理器的任何类型仪器有关的保护问题的许多解决办法。

（2）第二章提供与任何类型仪器有关的保护问题的许多保护办法，这类仪器使用微处理器，这些微处理器当供电电压波动时必须予以保护。

（3）第二章提供解决某些保护问题的许多办法，这些保护问题与任何使用这样一些微处理器的仪器有关：当供电电压波动时时，必须予以保护。

例 5. This is a software <u>of flight simulator running on personal computers, intended primarily for computer-aided instruction which requires providing practice navigation and instruments reading while users immerse in the virtual environment.</u>

这个句子中的修饰关系也是"连环式"。由于某些修饰环（如 of …和 running …, intended …和 which …）扣得紧，加上 which 引导的定语从句本身含有一个 while 引导的状语从句，所以此句也适于采用"解环法+先提后叙法"。全句可译为：

这是一个运行于个人计算机的飞行模拟器软件，这种软件主要用于如下计算机辅助教学：当用户投入虚拟环境时，它必须提供导航和仪表读数。

由此可见，翻译"连环式"修饰成分要因句制宜，灵活处理，不能只着眼于表面的修饰关系，还要考虑"修饰环"之间的紧密程度。通常，如果"修饰环"彼此关系较松，适于采用"解环法"；要是"扣"得很紧，则可采用"解环法+先提后叙法"。

四、句子结构调整法

有时，原文句中一个中心词带有若干修饰成分，但它们既不是纯"并列"头系，也不是规则的"连环"关系，而是"并列"中有"连环"（例1），或"连环"中含不规则"修饰环"（例2）；或者句中各有一个分别被"连环式"定语和"并列式"定语所修饰的中心词（例3）；或者中心词的定语从句本身又含有其他修饰成分（例 4）。由于这类句子所含的修饰关系比较复杂，很难纳入上述几种译法予以表达，此时就应该根据上下文的逻辑关系，调整句子结构。常用的方法是："化整为零"，将带有多重定语（从句）的长句拆译成若干汉语短句，然后按汉语表达习惯组织译文句子。例如：

例 1. An operating system is a master control program, permanently stored in memory, that interprets user commands requesting various kind of service, such as display, print, or copy a data file, list all files in a directory, or execute a particular program.

译文：操作系统是主控程序，永久地驻留在内存中，它理解用户的各种指令：如显示、打印文件，将目录中所有文件列表，或者执行一个特殊的程序。

（比较：操作系统是永久地驻留在内存中的；理解用户的各种指令：如显示、打印文件，将目录中所有文件列表，或者执行一个特殊的程序主控程序。）

例 2. The users of such a system control the process by means of a program, which is a set of instruction that specify the operation, operands, and the sequence by which processing has to occur.

译文：该系统用户通过程序控制处理过程，所谓程序是一套指定操作、操作数和处理序列的指令集。

（比较：该系统用户通过一套指定操作、操作数和处理序列的指令集即程序控制处理过程。）

例 3. Indeed, today's products — most of all, the latest in speech recognition – are a roll-out of technologies that have been percolating for years and that are based on an understanding of speech that has taken several decades to accumulate.

译文：的确，今日的产品，尤其是语音识别方面最新的产品，是过去多年技术渗透的延伸，也是基于几十年来对语音理解的结果。

（比较：的确，今日的产品，尤其是语音识别方面最新的产品，是过去多年的渗透和基于几十年来对语音理解的积累的技术延伸。）

例 4. The computer family, in computer science, is a term commonly used to indicate a group of computers that are built around the same microprocessor or around a series of related microprocessors, and that share significant design feature.

译文：在计算机科学中，计算机系列是常用的一个术语，通常指一组用相同的或者一系列相关的微处理器制造的计算机。

（比较：在计算机科学中，计算机系列是一个常用来指一组用相同的或者一系列相关的微处理器制造的计算机的术语。）

了解科技英语中复杂定语或长定语从句的译法，对提高科技英语翻译水平和译文质量具有重要意义。由于语言现象丰富多彩，修饰关系错综复杂，上述几种译法虽然适用于多数复杂定语或长定语从句，但不能包括一切场合，读者应根据特定的原文句子，灵活处理。

Exercises

I. Label each of the following statements as either true or false

1. The computer hardware recognizes only assembly language instruction.
2. A program written in the assembly language of one microprocessor can run on a computer that has a different microprocessor.
3. 4GLs are nonprocedural.
4. Each assembly language command corresponds to one unique machine-language command.
5. The lowest level of programming languages is machine language.

II. Compare the difference between machine languages, assembly languages and high-level languages

III. Describe each of the following terms

1. Programming Language
2. OOP
3. UML

IV. Questions

In what sense is a program in a third-generation language machine independent?
In what sense is it still machine dependent?

Chapter 8 The Internet

学习指导

Internet 是由一些使用公用语言互相通信的计算机连接而成的全球网络。Internet 与国际电话系统十分相似——没有人能完全拥有或控制它，但连接以后却能使它像大型网络一样运转。本章主要介绍 Internet 的关键技术。通过学习，读者应当：
- 掌握 Internet 的关键技术概念；
- 能够描述 Internet 协议与应用程序的作用；
- 掌握英语长句的翻译技巧。

What is the Internet? Where did it come from, and how did it support the growth of the World Wide Web? What are the Internet's most important operating principles? The **Internet** is an interconnected network of thousands of networks and millions of computers (sometimes called *host computers* or just *hosts*) linking businesses, educational institutions, government agencies, and individuals together. The Internet provides around 400 million people around the world (and over 170 million people in the United States) with services such as e-mail, newsgroups, shopping, research, instant messaging, music, videos, and news. No one organization controls the Internet or how it functions, nor is it owned by anybody, yet it has provided the infrastructure for a transformation in commerce, scientific research, and culture. The word *Internet* is derived from the word *internetwork* or the connecting together of two or more computer networks. The **World Wide Web**, or **Web** for short, is one of the Internet's most popular services, providing access to over one billion Web pages, which are documents created in a programming language called HTML and which can contain text, graphics, audio, video, and other objects, as well as "hyperlinks" that permit a user to jump easily from one page to another[1].

8.1 The Internet: Key Technology Concepts

In 1995, the Federal Networking Council (FNC) took the step of passing a resolution formally defining the term *Internet*.

"Internet" refers to the global information system that—
(i) is logically linked together by a globally unique address space based on the Internet Protocol (IP) or its subsequent extensions/follow-ons;
(ii) is able to support communications using the Transmission Control Protocol/Internet Protocol (TCP/IP) suite or its subsequent extensions/follow-ons, and/or other

IP-compatible protocols; and

(iii) *provides, uses or makes accessible, either publicly or privately, high level services layered on the communications and related infrastructure described herein.*

Based on the definition, the Internet means a network that uses the IP addressing scheme, supports the Transmission Control Protocol (TCP), and makes services available to users much like a telephone system makes voice and data services available to the public[2].

Behind this formal definition are three extremely important concepts that are the basis for understanding the Internet: packet switching, the TCP/IP communications protocol, and client/server computing[3]. Although the Internet has evolved and changed dramatically in the last 30 years, these three concepts are at the core of how the Internet functions today and are the foundation for Internet tomorrow.

Packet Switching. Packet switching is a method of slicing digital messages into parcels called "**packets**," sending the packets along different communication paths as they become available, and then reassembling the packets once they arrive at their destination. Prior to the development of packet switching, early computer networks used leased, dedicated telephone circuits to communicate with terminals and other computers. In circuit-switched networks such as the telephone system, a complete point-to-point circuit is put together, and then communication can proceed. However, these "dedicated" circuit-switching techniques were expensive and wasted available communications capacity—the circuit would be maintained regardless of whether any data was being sent. For nearly 70% of the time, a dedicated voice circuit is not being fully used because of pauses between words and delays in assembling the circuit segments, both of which increased the length of time required to find and connect circuits. A better technology was needed.

The first book on packet switching was written by Leonard Kleinrock in 1964, and the technique was further developed by others in the defense research labs of both the United States and England. With packet switching, the communications capacity of a network can be increased by a factor of 100 or more. The communications capacity of a digital network is measured in terms of bits per second. Imagine if the gas mileage of your car went from 15 miles per gallon to 1,500 miles per gallon—all without changing too much of the car!

In packet-switched networks, messages are first broken down into packets. Appended to each packet are digital codes that indicate a source address (the origination point) and a destination address, as well as sequencing information and error control information for the packet. Rather than being sent directly to the destination address, in a packet network, the packets travel from computer to computer until they reach their destination. These computers are called routers. **Routers** are special purpose computers that interconnect the thousands of different computer networks that make up the Internet and route packets along to their ultimate destination as they travel[4]. To ensure that packets take the best available path toward their destination, the routers use computer programs called **routing algorithms**.

Packet switching does not require a dedicated circuit but can make use of any spare capacity that is available on any of several hundred circuits. Packet switching makes nearly full use of almost all available communication lines and capacity. Moreover, if some lines are disabled or too busy, the packets can be sent on any available line that eventually leads to the destination point.

TCP/IP. While packet switching was an enormous advance in communications capacity, there was no universally agreed upon method for breaking up digital messages into packets, routing them to the proper address, and then reassembling them into a coherent message. This was like having a system for producing stamps, but no postal system (a series of post offices and a set of addresses).

TCP/IP answered the problem of what to do with packets on the Internet and how to handle them. **TCP** refers to the Transmission Control Protocol (TCP). **IP** refers to the Internet Protocol (IP). A **protocol** is a set of rules for formatting, ordering, compressing, and error-checking messages. It may also specify the speed of transmission and means by which devices on the network will indicate they have stopped sending and/or receiving messages. Protocols can be implemented in either hardware or software. TCP/IP is implemented in Web software called *server software* (described below). TCP is the agreed upon protocol for transmitting data packets over the Web. TCP establishes the connections among sending and receiving Web computers, handles the assembly of packets at the point of transmission, and their reassembly at the receiving end.

TCP/IP is divided into four separate layers, with each layer handling a different aspect of the communication problem. The Network Interface Layer is responsible for placing packets on and receiving them from the network medium, which could be a Local Area Network (Ethernet) or Token Ring Network, or other network technology. TCP/IP is independent from any local network technology and can adapt to changes in the local level. The Internet Layer is responsible for addressing, packaging, and routing messages on the Internet. The Transport Layer is responsible for providing communication with the application by acknowledging and sequencing the packets to and from the application. The Application Layer provides a wide variety of applications with the ability to access the services of the lower layers. Some of the best known applications are Hyper Text Transfer Protocol (HTTP), File Transfer Protocol (FTP), and Simple Mail Transfer Protocol (SMTP), all of which we will discuss later in this chapter.

IP Addresses. TCP handles the packetizing and routing of Internet messages. IP provides the Internet's addressing scheme. Every computer connected to the Internet must be assigned an address—otherwise it cannot send or receive TCP packets. For instance, when you sign onto the Internet using a dial-up telephone modem, your computer is assigned a temporary address by your Internet Service Provider.

Internet addresses, known as **IP addresses**, are 32-bit numbers that appear as a series of four separate numbers marked off by periods, such as 201.61.186.227. Each of the four numbers can range from 0–255. This "dotted quad" addressing scheme contains up to 4 billion addresses (2^{32}). The leftmost number typically indicates the network address of the computer, while remaining numbers help to identify the specific computer within the group that is sending (or receiving) a message.

The current version of IP is called Version 4, or IPv4. Because many large corporate and government domains have been given millions of IP addresses each (to accommodate their current and future work forces), and with all the new networks and new Internet-enabled devices requiring unique IP addresses being attached to the Internet, a new version of the IP protocol, called IPv6 is being adopted. This scheme contains 128-bit addresses, or about one quadrillion (10^{15}).

Figure 8-1 illustrates how TCP/IP and packet switching work together to send data over the Internet.

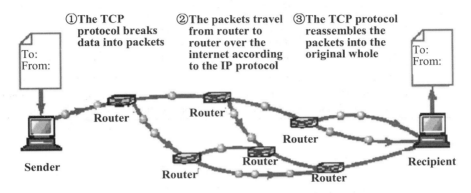

Fig.8-1　Routing Internet messages: TCP/IP and packet switching

Domain Names and URLs. Most people cannot remember 32-bit numbers. IP addresses can be represented by a natural language convention called **domain names. The domain name system (DNS)** allows expressions such as cnet.com to stand for numeric IP addresses (cnet.com's numeric IP is 216.200.247.134).**Uniform resource locators (URLs)**, which are the addresses used by Web browsers to identify the location of content on the Web, also use domain names as part of the URL. A typical URL contains the protocol to be used when accessing the address, followed by its location. For instance, the URL htttp://www.azimuth-interactive.com/flash_test refers to the IP address 208.148.84.1 with the domain name "azimuth-interactive.com" and the protocol being used to access the address, Hypertext Transfer Protocol (HTTP). A resource called "flash_test" is located on the server directory path /flash_test. A URL can have from two to four parts, for example name1.name2.name3.org. We discuss domain names and URLs further in next chapter. Table 8-1 summarizes the important components of the Internet addressing scheme.

Client/Server Computing. While packet switching exploded the available communications capacity and TCP/IP provided the communications rules and regulations, it took a revolution in computing to bring about today's Internet and the Web. That revolution is called *client/server computing* and without it, the Web—in all its richness—would not exist. In fact, the Internet is a giant example of client/server computing in which over 70 million host server computers store Web pages and other content that can be easily accessed by nearly a million local area networks and hundreds of millions of client machines worldwide[5].

Table 8-1 Pieces of the Internet puzzle: names and addresses

IP addresses	Every computer connected to the Internet must have a unique address number called an *Internet Protocol address*. Even computers using a modem are assigned a temporary IP address.
Domain names	The DNS (domain name system) allows expressions such as aw.com (Addison Wesley's Web site) to stand for numeric IP locations.
DNS servers	DNS servers are databases that keep track of IP addresses and domain names on the Internet.
Root servers	Root servers are central directories that list all domain names currently in use. DNS servers consult root servers to look up unfamiliar domain names when routing traffic.
ICANN	The Internet Corporation for Assigned Numbers and Names (ICANN) was established in 1998 to set the rules for domain names and IP addresses and also to coordinate the operation of root servers. It took over from private firms such as NetSolutions.com.

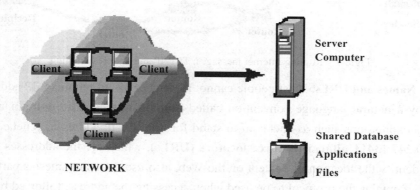

Fig 8-2 the client/server computing model

Client/server computing is a model of computing in which very powerful personal computers called **clients** are connected together in a network together with one or more server computers. These clients are sufficiently powerful to accomplish complex tasks such as displaying rich graphics, storing large files, and processing graphics and sound files, all on a local desktop or handheld device. **Servers** are networked computers dedicated to common functions that the client machines on the network need, such as storing files, software applications, utility programs such as Web connections, and printers (See Figure 8-2).

To appreciate what client/server computing makes possible, you must understand what preceded it. In the mainframe computing environment of the 1960s and 1970s, computing power was very expensive and limited. For instance, the largest commercial mainframes of the late 1960s had 128k of RAM and 10 megabyte disk drives, and occupied hundreds of square feet. There was insufficient computing capacity to support graphics or color in text documents, let alone sound files or hyper linked documents and databases.

With the development of personal computers and local area networks during the late 1970s and early 1980s, client/server computing became possible. Client/server computing has many advantages over centralized mainframe computing. For instance, it is easy to expand capacity by adding servers and clients. Also, client/server networks are less vulnerable than centralized computing architectures. If one server goes down, backup or mirror servers can pick up the slack; if a client machine is inoperable, the rest of the network continues operating. Moreover, processing load is balanced over many powerful smaller machines rather than being concentrated in a single huge machine that performs processing for everyone. Both software and hardware in client/server environments can be built more simply and economically.

New Words & Expressions

hyperlink   n. [计]超链接
video   adj. n. 视频
megabyte   n. [计]兆字节
quad   n. 四元组
vulnerable   adj. 易受攻击的，易受…的攻击
gigabyte   n. 十亿字节（giga-为字首，"十亿"）
host computer (host)   主机
packet-switching   分组交换技术
circuit switching   线路转接
Router   [计] 路由器
Ethernet   n. 以太网

audio   adj. n. 音频
quadrillion   n. 千的五次方，adj. 千之五次方的
protocol   n. 草案，协议
mainframe   n. [计]主机，大型机
Domain Name   域名
supercomputer   n. [计]超级计算机
instant messaging   即时信息服务
client/server   客户机/服务器
bit   n. [计]位，比特
routing algorithm   路径算法
backup   vt. 做备份；adj. [计]文件备份

Abbreviations

TCP (Transfer Control Protocol) 传输控制协议
URL (Uniform Resource Locator) 统一资源定位
RAM (random access memory) 随机存储器
ICANN (Internet Corporation for Assigned Names and Numbers) 因特网域名与地址管理组织
IP(Internet Protocol) 网际协议
DNS (domain name system) 域名系统
IP address   IP 地址

Notes:

1. The **World Wide Web**, or **Web** for short, is …, providing access to over one billion Web pages, which are … and which can contain … as well as "hyperlinks" that… another. "全球信息网，或简称万维网，是英特网上最流行的服务之一，提供对超过 10 亿网页的访问，这些网页是由一种叫做 HTML （超文本链接标示语言）的编程语言生成的文件，它可以包含本文、图形、声频、视频和其他对象、以及允许用户容易地跳跃到其他网页的'超链接'。"

2. Based on the definition, the Internet means a network that uses …, supports …, and makes …. 本句 that 引起一个由三个并列句组成的定语从句，译为："基于这个定义，因特网表示这样一个网络——使用 IP 地址分配方案、支持传输控制协议，而且非常类似于电话系统使公众能够使用声音和数据服务它使用户可以使用多种服务。"

3. Behind this formal definition are three extremely important concepts that are the basis for understanding

the Internet: packet switching, the TCP/IP communications protocol, and client/server computing. 本句是倒状句，译为："在这个正式的定义背后，隐含着三个极端重要的概念：分组交换、TCP/IP（传输控制协议/网际协议）通信协议和客户机/服务器计算技术，它们乃是理解因特网的基础。"

4. **Routers** are special purpose computers that interconnect the thousands of different computer networks that make up the Internet and route packets along to their ultimate destination as they travel. 本句中第一个 that 引起的定语从句是并列句，interconnect，and route 为两个动词。译为："路由器是一种特殊用途的计算机，它将组成因特网的成千上万个不同计算机网络互相联接起来，并在信息包旅行时将它们向终极目的地发送。"

5. In fact, the Internet is a giant example of client/server computing in which over 70 million host server computers store Web pages and other content that can be easily accessed by nearly a million local area networks and hundreds of millions of client machines worldwide. "事实上，因特网是客户机/服务器计算技术的一个巨大实例，其中，超过 7000 万部主机服务器计算机储存网页和其他内容，这些网页和内容能被全世界接近一百万个局域网和数亿台客户机容易地访问。"

8.2 Other Internet Protocols and Utility Programs

There are many other Internet protocols that provide services to users in the form of Internet applications that run on Internet clients and servers. These Internet services are based on universally accepted protocols—or standards—that are available to everyone who uses the Internet. They are not owned by any one organization but are services that were developed over many years and given to all Internet users.

HTTP: Hypertext Documents. HTTP (short for **Hyper Text Transfer Protocol**) is the Internet protocol used for transferring Web pages (described in the following section). The HTTP protocol runs in the Application Layer of the TCP/IP model. An HTTP session begins when a client's browser requests a Web page from a remote Internet server. When the server responds by sending the page requested, the HTTP session for that object ends. Because Web pages may have many objects on them—graphics, sound or video files, frames, and so forth—each object must be requested by a separate HTTP message.

SMTP, POP, and IMAP: Sending E-mail. E-mail is one of the oldest, most important, and frequently used Internet services. **STMP (Simple Mail Transfer Protocol)** is the Internet protocol used to send mail to a server. **POP (Post Office Protocol)** is used by the client to retrieve mail from an Internet server. You can see how your browser handles SMTP and POP by looking in your browser's Preferences or Tools section, where the mail settings are defined. You can set POP to retrieve e-mail messages from the server and then delete the messages on the server, or retain them on the server. **IMAP (Internet Message Access Protocol)** is a more current e-mail protocol supported by many servers and all browsers. IMAP allows users to search, organize, and filter their mail prior to downloading it from the server.

FTP: Transferring Files. FTP (**File Transfer Protocol**) is one of the original Internet services. It is a part of the TCP/IP protocol and permits users to transfer files from the server to their client machine, and vice versa. The files can be documents, programs, or large database files. FTP is the fastest and most convenient way to transfer files larger than 1 megabyte, which many mail servers will not accept.

SSL: Security. SSL (Secure Sockets Layer) is a protocol that operates between the Transport and Application Layers of TCP/IP and secures communications between the client and the server. SSL helps secure e-commerce communications and payments through a variety of techniques such as message encryption and digital signatures.

Telnet: Running Remote. Telnet is a terminal emulation program that runs in TCP/IP. You can run Telnet from your client machine. When you do so, your client emulates a mainframe computer terminal. (The industry standard terminals defined in the days of mainframe computing are VT-52, VT-100, and IBM 3250.) You can then attach yourself to a computer on the Internet that supports Telnet and run programs or download files from that computer. Telnet was the first "remote work" program that permitted users to work on a computer from a remote location.

Finger: Finding People. You can find out who is logged onto a remote network by using Telnet to connect to a server, and then typing "finger" at the prompt. **Finger** is a utility program supported by UNIX computers. When supported by remote computers, finger can tell you who is logged in, how long they have been attached, and their user name. Obviously there are security issues involved with supporting finger, and most Internet host computers do not support finger today.

Ping: Testing the Address. You can "**ping**" a host computer to check the connection between your client and the server. The ping (Packet InterNet Groper) program will also tell you the time it takes for the server to respond, giving you some idea about the speed of the server and the Internet at that moment. You can run ping from the DOS prompt on a personal computer with a Windows operating system by typing: Ping <domain name>.

Tracert: Checking Routes. Tracert is one of a several route-tracing utilities that allow you to follow the path of a message you send from your client to a remote computer on the Internet.

New Words & Expressions

Telnet 远程登录
Ping 测试IP地址的程序
Finger 查找器（查找因特网用户的程序）
Tracert 检查路由器程序

Abbreviations

FTP (File Transfer Protocol) 文件传输协议
SSL (Security Socket Layer) 加密套接字协议层
SMTP (Simple Message Transfer Protocol) 简单邮件传输协议，用于电子邮件的传输
IMAP (Internet Message Access Protocol) Internet消息访问协议
POP (Post Office Protocol) 邮局协议
RAM (random access memory) 随机存储器

8.3 Internet Service Providers

The firms that provide the lowest level of service in the multi-tiered Internet architecture by leasing Internet access to home owners, small businesses, and some large institutions are called **Internet Service Providers (ISPs)**[1]. ISPs are retail providers—they deal with "the last mile of service" to the curb, the home, the business office. About 45 million American households connect to the Internet through either national or local ISPs. ISPs typically connect to the Internet and MAEs or NAPs with high-speed telephone or cable lines (up to 45 Mbps).

There are major ISPs such as America Online, MSN Network, and AT&T World-Net and about 5,000 local ISPs in the United States, ranging from local telephone companies offering dial-up and DSL telephone access, to cable companies offering cable modem service, to small "mom and pop" Internet shops that service a small town, city, or even county with mostly dial-up phone access. If you have home or small business Internet access, an ISP will be providing you the service.

There are two types of ISP service: narrowband and broadband. **Narrowband** service is the traditional telephone modem connection now operating at 56.6 Kbps (although the actual throughput hovers around 30 Kbps due to line noise that causes extensive resending of packets). This is the most common form of connection worldwide. **Broadband** service is based on DSL, cable modem, telephone (T1 and T3 lines), and satellite technologies. Broadband—in the context of Internet service—refers to any communication technology that permits clients to play streaming audio and video files at acceptable speeds—generally anything above 100 Kbps.

The term **DSL** refers to **digital subscriber line** service, which is a telephone technology for delivering high-speed access through ordinary telephone lines found in your home or business. Service levels range from about 150 Kbps all the way up to 1 Mbps. DSL service requires that customers live within two miles (about 4,000 meters) of a neighborhood telephone switching center.

Cable modem refers to a cable television technology that piggybacks digital access to the Internet on top of the analog video cable providing television signals to a home[2]. Cable modem services ranges from 350 Kbps up to 1 Mbps. Cable service may degrade if many people in a neighborhood log on and demand high-speed service all at once.

T1 and **T3** are international telephone standards for digital communication. T1 lines offer guaranteed delivery at 1.54 Mbps, while T3 lines offer delivery at 43 Mbps. T1 lines cost about $1,000–$2,000 per month, and T3 lines between $10,000 and $30,000 per month. These are leased, dedicated, guaranteed lines suitable for corporations, government agencies, and businesses such as ISPs requiring high-speed guaranteed service levels.

Some satellite companies are offering broadband high-speed digital downloading of Internet content to homes and offices that deploy 18″ satellite antennas. Service is available beginning at 256 Kbps up to 1 Mbps. In general, satellite connections are not viable for homes and small businesses because they are only one-way—you can download from the Internet at high speed, but cannot upload to the Internet at all. Instead, users require a phone or cable connection for their uploading.

New Words & Expressions

tier n. 一层，一排   curb n. 路边

mom and pop shop　n. 小零售铺，夫妻店
narrowband　n. 窄带
throughput　n. 吞吐量，生产能力
modem　n. 调制解调器
aficionado　n. <西班牙>狂热爱好者，迷
MBPS　n. [计]兆比特每秒
lengthy　adj. 冗长的，过分的

dial-up　拨号上网
broadband　n. 宽带
objectionable　adj. 该反对的，不能采用的
delay　n. 延迟
sovereign　adj. 统治的；n. 统治者
piggyback　v. 搭载

Abbreviations

NAP (Network Access Point)　网络访问节点
DSL (Digital Subscriber Line)　数字用户线
MAE (Metropolitan Area Exchange)　城域交换站
ISP (Internet Service Provider) Internet 服务提供者

Notes

1. The firms that provide the lowest level of service in the multi-tiered Internet architecture by leasing Internet access to home owners, small businesses, and some large institutions are called Internet Service Providers (ISPs). "在多层次因特网体系中通过向家庭出租因特网访问通道提供最低水平服务的公司、小型企业，和一些大机构叫做因特网服务提供商（ISP）。"

2. Cable modem refers to a cable television technology that piggybacks digital access to the Internet on top of the analog video cable providing television signals to a home. "电缆调制解调器是一种在向家庭提供电视信号的视频电缆上搭载对因特网进行数字式访问的有线电视技术。"

Reading Material: Transition from IPv4 to IPv6

The most important idea behind the transition from IPv4 to IPv6 is that the Internet is far too big and decentralized to have a "flag day"—one specified day on which every host and router is upgraded from IPv4 to IPv6. Thus, IPv6 needs to be deployed incrementally in such a way that hosts and routers that only understand IPv4 can continue to function for as long as possible. Ideally, IPv4 nodes should be able to talk to other IPv4 nodes and some set of other IPv6-capable nodes indefinitely. Also, IPv6 hosts should be capable of talking to other IPv6 nodes even when some of the infrastructure between them may only support IPv4. Two major mechanisms have been defined to help this transition: dual-stack operation and tunneling.

The idea of dual stacks is fairly straightforward: IPv6 nodes run both IPv6 and IPv4 and use the Version field to decide which stack should process an arriving packet. In this case, the IPv6 address could be unrelated to the IPv4 address, or it could be the IPv4-mapped IPv6 address described earlier in this section.

The basic tunneling technique, in which an IP packet is sent as the payload of another IP packet, was described in Section 8.1. For IPv6 transition, tunneling is used to send an IPv6 packet over a piece of the network that only understands IPv4. This means that the IPv6 packet is encapsulated within an IPv4 header that has the address of the tunnel endpoint in its header, is transmitted across the IPv4-only piece of network, and then is decapsulated at the endpoint. The endpoint could be either a router or a host; in either case, it must be IPv6 capable to be able to process the IPv6 packet after decapsulation. If the endpoint is a host with an IPv4-mapped IPv6 address, then tunneling can be done automatically by extracting the IPv4 address from the IPv6 address and using it to form the IPv4 header. Otherwise, the tunnel must be configured manually. In this case, the encapsulating node needs to know the IPv4 address of the other end of the tunnel, since it cannot be extracted from the IPv6 header. From the perspective of IPv6, the other end of the tunnel looks like a regular IPv6 node that is just one hop away, even though there may be many hops of IPv4 infrastructure between the tunnel endpoints.

New Words & Expressions

decentralize　vt. 分散　n. 分散　　　　　indefinitely　adv. 无限期地
encapsulated　adj. 封装的　　　　　　　router　n. 路由器
host　n. 主计算机，主机

英语长句的翻译

在翻译英语长句时，首先，不要因为句子太长而产生畏惧心理，因为，无论是多么复杂的句子，它都是由一些基本的成分组成的。其次要弄清英语原文的句法结构，找出整个句子的中心内容及其各层意思，然后分析几层意思之间的相互逻辑关系，再按照汉语的特点和表达方式，正确地译出原文的意思，不必拘泥于原文的形式。

一、英语长句的分析

一般来说，造成长句的原因有三方面：
（1）修饰语过多；
（2）并列成分多；
（3）语言结构层次多。
在分析长句时可以采用下面的方法：
（1）找出全句的主语、谓语和宾语，从整体上把握句子的结构。
（2）找出句中所有的谓语结构、非谓语动词、介词短语和从句的引导词。
（3）分析从句和短语的功能，例如，是否为主语从句，宾语从句，表语从句等，若是状语，它是表示时间、原因、结果、还是表示条件等等。
（4）分析词、短语和从句之间的相互关系，例如，定语从句所修饰的先行词是哪一个等。
（5）注意插入语等其他成分。
（6）注意分析句子中是否有固定词组或固定搭配。
下面我们结合一些实例来进行分析：
例 1. Behaviorists suggest that the child who is raised in an environment where there are many stimuli which develop his or her capacity for appropriate responses will experience greater intellectual development.

分析：①该句的主语为 behaviorists，谓语为 suggest，宾语为一个从句，因此整个句子为 Behaviorist suggest that-clause 结构。②该句共有五个谓语结构，它们的谓语动词分别为 suggest, is raised, are, develop, experience 等，这五个谓语结构之间的关系为：Behaviorist suggest that-clause 结构为主句；who is raised in an environment 为定语从句，所修饰的先行词为 child；where there are many stimuli 为定语从句，所修饰的先行词为 environment; which develop his or her capacity for appropriate responses 为定语从句，所修饰的先行词为 stimuli；在 suggest 的宾语从句中，主语为 child，谓语为 experience，宾语为 greater intellectual development。

在作了如上的分析之后，我们就会对该句具有了一个较为透彻的理解，然后根据我们上面所讲述的各种翻译方法，就可以把该句翻译成汉语为：

行为主义者认为，如果儿童的成长环境里有许多刺激因素，这些因素又有利于其适当反应能力的发展，那么，儿童的智力就会发展到较高的水平。

例 2. For a family of four, for example, it is more convenient as well as cheaper to sit comfortably at home, with almost unlimited entertainment available, than to go out in search of amusement elsewhere.

分析：①该句的骨干结构为 it is more … to do sth than to do sth else.是一个比较结构，而且是在两个不定式之间进行比较。②该句中共有三个谓语结构，它们之间的关系为：it is more convenient as well as cheaper to …为主体结构，但 it 是形式主语，真正的主语为第二个谓语结构：to sit comfortably at home，并与第三个谓语结构 to go out in search of amusement elsewhere 作比较。③句首的 for a family of four 作状语，表示条件。另外，还有两个介词短语作插入语：for example，with almost unlimited entertainment available，其中第二个介词短语作伴随状语，修饰 to sit comfortably at home.

综合上述翻译方法，这个句子我们可以翻译为：

譬如，对于一个四口之家来说，舒舒服服地在家中看电视，就能看到几乎数不清的娱乐节目，这比到外面别的地方去消遣又便宜又方便。

二、长句的翻译

英语习惯于用长的句子表达比较复杂的概念，而汉语则不同，常常使用若干短句，作层次分明的叙述。因此，在进行英译汉时，要特别注意英语和汉语之间的差异，将英语的长句分解，翻译成汉语的短句。在英语长句的翻译过程中，我们一般采取下列的方法。

2.1 顺序法

当英语长句内容的叙述层次与汉语基本一致时，可以按照英语原文的顺序翻译成汉语。

例 1. Even when we turn off the beside lamp and are fast asleep, electricity is working for us, driving our refrigerators, heating our water, or keeping our rooms air-conditioned.

分析：该句子由一个主句，三个作伴随状语的现在分词以及位于句首的时间状语从句组成，共有五层意思：①既使在我们关掉了床头灯深深地进入梦乡时；②电仍在为我们工作；③帮我们开动电冰箱；④加热水；⑤或是室内空调机继续运转。上述五层意思的逻辑关系以及表达的顺序与汉语完全一致，因此，我们可以通过顺序法，把该句翻译成：

即使在我们关掉了床头灯深深地进入梦乡时，电仍在为我们工作：帮我们开动电冰箱，把水加热，或使室内空调机继续运转。

例 2. But now it is realized that supplies of some of them are limited, and it is even possible to give a reasonable estimate of their "expectation of life", the time it will take to exhaust all known sources and reserves of these materials.

分析：该句的骨干结构为"It is realized that…"，it 为形式主语，that 引导着主语从句以及并列的 it is even possible to …结构，其中，不定式作主语，the time …是"expectation of life"的同位语，进一步解释其含义，而 time 后面的句子是它的定语从句。五个谓语结构，表达了四个层次的意义：①可是现在人们意识到；②其中有些矿物质的蕴藏量是有限的；③人们甚至还可以比较合理的估计出这些矿物质"可望存在多少年"；④将这些已知矿源和储量将消耗殆尽的时间。根据同位语从句的翻译方法，把第四层意义的表达作适当的调整，整个句子就翻译为：

可是现在人们意识到，其中有些矿物质的蕴藏量是有限的，人们甚至还可以比较合理的估计出这些矿物质"可望存在多少年"，也就是说，经过若干年后，这些矿物的全部已知矿源和储量将消耗殆尽。

2.2 逆序法

英语有些长句的表达次序与汉语表达习惯不同，甚至完全相反，这时必须从原文后面开始翻译。例如：

例 1. Aluminum remained unknown until the nineteenth century, because nowhere in nature is it found free, owing to its always being combined with other elements, most commonly with oxygen, for which it has a strong affinity.

分析：这个句子由一个主句，两个原因状语和一个定语从句，"铝直到 19 世纪才被人发现"是主句，也是全句的中心内容，全句共有四个谓语结构，共有五层意思：①铝直到 19 世纪才被人发现；②由于在自然界找不到游离状态的铝；③由于它总是跟其他元素结合在一起；④最普遍的是跟氧结合；⑤铝跟氧有很强的亲和力。按照汉语的表达习惯通常因在前，果在后，这样，我们可以逆着原文的顺序把该句翻译成：

铝总是跟其他元素结合在一起，最普遍的是跟氧结合；因为铝跟氧有很强的亲和力，由于这个原因，在自然界找不到游离状态的铝。所以，铝直到 19 世纪才被人发现。

例 2. It therefore becomes more and more important that, if students are not to waste their opportunities, there will have to be much more detailed information about courses and more advice.

分析：该句由一个主句，一个条件状语从句和一个宾语从句组成，"……变得越来越重要"是主句，也是全句的中心内容，全句共有三个谓语结构，包含三层含义：①…变的越来越重要；②如果要使学生充分利用他们的机会；③得为他们提供大量更为详尽的信息，作更多的指导。为了使译文符合汉语的表达习惯，我们也采用逆序法，翻译成：

因此，如果要使学生充分利用他们（上大学）的机会，就得为他们提供大量关于课程的更详尽信息，作更多的指导。这个问题显得越来越重要了。

例 3. It is probably easier for teachers than for students to appreciate the reasons why learning English seems to become increasingly difficult once the basic structures and patterns of the language have been understood.

一旦了解英语的基本结构和句型，再往下学似乎就越来越难了，这其中的原因，也许教师比学生更容易理解。

例 4. For our purposes we will say e-commerce begins in 1995, following the appearance of the first banner advertisements placed by ATT, Volvo, Sprint and others on Hotwired.com in late October 1994, and the first sales of banner ad space by Netscape and Infoseek in early 1995.

伴随着 ATT、Volvo、Sprint 等公司所做的第一例横幅广告于 1994 年 10 月下旬出现在 Hotwired.com 上，和 1995 年初 Netscape 与 Infoseek 领先出售横幅广告空间，我们会说电子商务是从 1995 年开始的。

2.3 分句法

有时英语长句中主语或主句与修饰词的关系并不十分密切，翻译时可以按照汉语多用短句的习惯，把长句的从句或短语化成句子，分开来叙述，为了使语意连贯，有时需要适当增加词语。例如：

例 1. The number of the young people in the United States who can't read is incredible about one in four.

上句在英语中是一个相对简单的句子，但是如果我们按照原文的句子结构死译，就可能被翻译成："没有阅读能力的美国青年人的数目令人难以置信约为 1/4。"这样，就使得译文极为不通顺，不符合汉语的表达习惯，因此，我们应该把它译为：

大约有 1/4 的美国青年人没有阅读能力，这简直令人难以置信。

例 2. Television, it is often said, keeps one informed about current events, allow one to follow the latest developments in science and politics, and offers an endless series of programs which are both instructive and entertaining.

分析：在此长句中，有一个插入语"it is often said"，三个并列的谓语结构，还有一个定语从句，这三个并列的谓语结构尽管在结构上同属于同一个句子，但都有独立的意义，因此在翻译时，可以采用分句法，按照汉语的习惯把整个句子分解成几个独立的分句，结果为：

人们常说，通过电视可以了解时事，掌握科学和政治的最新动态。从电视里还可以看到层出不穷、既有教育意义又有娱乐性的新节目。

例 3. All they have to do is press a button, and they can see plays, films, operas, and shows of every kind, not to mention political discussions and the latest exciting football match.

他们所必须做的只是按一下开关。开关一开，就可以看到电视剧、电影、歌剧、以及其他各种各样的文艺节目。至于政治问题的辩论、最近的激动人心的足球赛更是不在话下。

2.4 综合法

上面我们讲述了英语长句的逆序法、顺序法和分句法，事实上，在翻译一个英语长句时，并不只是单纯地使用一种翻译方法，而是要求我们把各种方法综合使用，这在我们上面所举的例子中也有所体现。尤其是在一些情况下，一些英语长句单纯采用上述任何一种方法都不方便，这就需要我们的仔细分析，或按照时间的先后，或按照逻辑顺序，顺逆结合，主次分明地对全句进行综合处理，以便把英语原文翻译成通顺忠实的汉语句子。例如：

例 1. People were afraid to leave their houses, for although the police had been ordered to stand by in case of emergency, they were just as confused and helpless as anybody else.

分析：该句共有三层含义：①人们不敢出门；②尽管警察已接到命令，要作好准备以应付紧急情况；③警察也和其他人一样不知所措和无能为力。在这三层含义中，②表示让步，③表示原因，而①则表示结果，按照汉语习惯顺序，我们作如下的安排：

尽管警察已接到命令，要作好准备以应付紧急情况，但人们不敢出门，因为警察也和其他人一样不知所措和无能为力。

例 2. Napster.com, which was established to aid Internet users in finding and sharing online music files known as *MP3 files*, is perhaps the most wellknown example of peer-to-peer e-commerce, although purists note that Napster is only partially peer-to-peer because it relies on a central database to show which users are sharing music files.

Napster.com 建立的目标是帮助因特网用户发现并分享在线音乐文件，即人所共知的 MP3 文件。尽管纯化论者强调：因为它依赖中央数据库来显示哪一位用户正在分享音乐文件，所以 Napster 仅仅是部分对等。但 Napster 或许是对等电子商务最著名的实例。

Exercises

I. Answer the following questions

1. Relate the key technology concepts behind the Internet.
2. Please describe the role of Internet protocols and utility programs.

II. Fill in the blanks in each of the following

1. The basic building blocks of the Internet are:_____.
2. _____ is a method of slicing digital messages into parcels called "packets," sending the packets along different communication paths as they become available, and then reassembling the packets once they arrive at their destination.
3. IP addresses are 32-bit numbers that appear as a series of _____ separate numbers marked off by periods.
4. IP addresses can be represented by a natural language convention called _____.

Chapter 9   The World Wide Web

学习指导

WWW 是 Internet 的多媒体信息查询工具，是 Internet 上发展最快和最广泛的服务。你可以将 WWW 视为 Internet 上一个大型图书馆，"Web 节点"就象图书馆中的一本本书，而"Web 页"则是书中的某一页。多个 Web 页合在一起便组成了一个 Web 节点。可以从一个特定的 Web 节点开始您的 Web 环游之旅。本章主要分析万维网（WWW）的现在与将来，以及它们如何工作、如何进化。通过学习，读者应该：

- 理解万维网（World Wide Web）的工作原理；
- 能够描述 Internet 和万维网的特征及其服务；
- 了解学术论文写作的有关知识。

The invention of the Web brought an extraordinary expansion of digital services to millions of amateur computer users, including color text and pages, formatted text, pictures, animations, video, and sound. In short, the Web makes nearly all the rich elements of human expression needed to establish a commercial marketplace available to nontechnical computer users worldwide.

9.1   Hypertext

Web pages can be accessed through the Internet because the Web browser software operating your PC can request Web pages stored on an Internet host server using the HTTP protocol. **Hypertext** is a way of formatting pages with embedded links that connect documents to one another, and that also link pages to other objects such as sound, video, or animation files. When you click on a graphic and a video clip plays, you have clicked on a hyperlink. For example, when you type a Web address in your browser such as http://www.sec.gov, your browser sends an HTTP request to the sec.gov server requesting the home page of sec.gov.

HTTP is the first set of letters at the start of every Web address, followed by the domain name. The domain name specifies the organization's server computer that is housing the document. Most companies have a domain name that is the same as or closely related to their official corporate name. The directory path and document name are two more pieces of information within the Web address that help the browser track down the requested page. Together, the address is called a Uniform Resource Locator, or URL. When typed into a browser, a URL tells it exactly where to look for the information. For example, in the following URL:

http://www.megacorp.com/content/features/082602.html

http=the protocol used to display Web pages;

www.megacorp.com = domain name;

content/features=the directory path that identifies where on the domain Web server the page is stored;

082602.html=document name and its format (an html page).

The most common domain extensions currently available and officially sanctioned by ICANN are shown in the list below. Countries also have domain names such as .uk, .au, and .fr (United Kingdom, Australia, and France). Also shown in the list below are recently approved top-level domains .biz and .info, as well as new domains under consideration. In the near future, this list will expand to include many more types of organizations and industries.

.com Commercial organizations/businesses

.edu Educational institutions

.gov U.S. government agencies

.mil U.S. military

.net Network computers

.org Nonprofit organizations and foundations

New Top-Level Domains approved May 15, 2001.

.biz business firms

.info information providers

New Top-Level Domains proposed.

.aero Air transport industry

.coop Cooperatives

.museum Museums

.name Individuals

.pro Professionals

New Words & Expressions

| animation | n. | 动画 | browse | v.n. | 浏览 |
| Hypertext | n. | 超文本 | protocol | n. | 草案，协议 |

9.2 Markup Languages

Although the most common Web page formatting language is HTML, the concept behind document formatting actually had its roots in the 1960s with the development of Generalized Markup Language (GML).

SGML. In 1986, the International Standards Organization adopted a variation of GML called **Standard Generalized Markup Language**, or **SGML**. The purpose of SGML was to help very large organizations format and categorize large collections of documents. The advantage of SGML is that it can run independent of any software program but, unfortunately, it is extremely complicated and difficult to learn. Probably for this reason, it has not been widely adopted.

HTML. HTML (**HyperText Markup Language**) is a GML that is relatively easy to use. HTML provides Web page designers with a fixed set of markup "tags" that are used to format a Web page. When these tags are inserted into a Web page, they are read by the browser and interpreted into a page display. You can see the source HTML code for any Web page by simply clicking on the "Page Source" command found in all browsers.

HTML functions to define the structure and style of a document, including the headings, graphic positioning, tables, and text formatting.5 Since its introduction, the two major browsers—Netscape's Navigator and Microsoft's Internet Explorer—have continuously added features to HTML to enable programmers to further refine their page layouts. Unfortunately, many of the enhancements only work in one company's browser, and this development threatens the attainment of a universal computing platform. Worse, building browsers with proprietary functionality adds to the costs of building e-commerce sites. Whenever you build an e-commerce site, special care must be taken to ensure the pages can be viewed by major browsers, even outdated versions of browsers.

HTML Web pages can be created with any text editor, such as Notepad or Wordpad, using Microsoft Word (simply save the Word document as a Web page) or any one of several Web page editors.

XML. Extensible Markup Language (XML) takes Web document formatting a giant leap forward. **XML** is a new markup language specification developed by the W3C (the World Wide Web Consortium). XML is a markup language like HTML, but it has very different purposes. Whereas the purpose of HTML is to control the "look and feel" and display of data on the Web page, XML is designed to describe data and information.

For instance, if you want to send a patient's medical record—including diagnosis, personal identity, medical history information, and any doctor's notes—from a database in Boston to a hospital in New York over the Web, it would be impossible using HTML. However, with XML, these rich documents (database records) for patients could be easily sent over the Web and displayed.

XML is "extensible," which means the tags used to describe and display data are defined by the user, whereas in HTML the tags are limited and predefined. XML can also transform information into new formats, such as by importing information from a database and displaying it as a table. With XML, information can be analyzed and displayed selectively, making it a more powerful alternative to HTML. This means that business firms, or entire industries, can describe all of their invoices, accounts payable, payroll records, and financial information using a Web-compatible markup language. Once described, these business documents can be stored on intranet Web servers and shared throughout the corporation.

XML is not yet a replacement for HTML. Currently, XML is fully supported only by Microsoft's Internet Explorer 5, and is not supported by Netscape (although this may change). Whether XML eventually supplants HTML as the standard Web formatting specification depends a lot on whether it is supported by future Web browsers. Currently, XML and HTML work side by side on the same Web pages. HTML is used to define how information should be formatted, and XML is being used to describe the data itself.

New Words & Expressions

extension   n. 扩展名
browser   n. 浏览器
supplant   vt. 排挤掉，代替
surf   vi. 作冲浪运动，vt. 在...冲浪

Abbreviations

GML (Generalized Markup Language) 通用置标语言
SGML (Standard Generalized Markup Language) 标准通用置标语言
XML (Extensible Markup Language) 可扩展链接标示语言

9.3 Web Servers and Clients

We have already described client/server computing and the revolution in computing architecture brought about by client/server computing. You already know that a server is a computer attached to a network that stores files, controls peripheral devices, interfaces with the outside world—including the Internet—and does some processing for other computers on the network.

But what is a Web server? **Web server software** refers to the software that enables a computer to deliver Web pages written in HTML to client machines on a network that request this service by sending an HTTP request. The two leading brands of Web server software are Apache, which is free Web server shareware that accounts for about 60% of the market, and Microsoft's NT Server software, which accounts for about 20% of the market.

Aside from responding to requests for Web pages, all Web servers provide some additional basic capabilities such as the following:

- *Security services*—These consist mainly of authentication services that verify that the person trying to access the site is authorized to do so. For Web sites that process payment transactions, the Web server also supports Secure Sockets Layer (SSL), the Internet protocol for transmitting and receiving information securely over the Internet. When private information such as names, phone numbers, addresses, and credit card data need to be provided to a Web site, the Web server uses SSL to ensure that the data passing back and forth from the browser to the server is not compromised.
- *File Transfer Protocol (FTP)*—This protocol allows users to transfer files to and from the server. Some sites limit file uploads to the Web server, while others restrict downloads, depending on the user's identity.
- *Search engine*—Just as search engine sites enable users to search the entire Web for particular documents, search engine modules within the basic Web server software package enable indexing of the site's Web pages and content, and permit easy keyword searching of the site's content. When conducting a search, a search engine makes use of an

index, which is a list of all the documents on the server. The search term is compared to the index to identify likely matches.

- ***Data capture***—Web servers are also helpful at monitoring site traffic, capturing information on who has visited a site, how long the user stayed there, the date and time of each visit, and which specific pages on the server were accessed. This information is compiled and saved in a log file, which can then be analyzed by a user log file. By analyzing a log file, a site manager can find out the total number of visitors, average length of each visit, and the most popular destinations, or Web pages.

The term *Web server* is sometimes also used to refer to the physical computer that runs Web server software. Leading manufacturers of Web server computers are IBM, Compaq, Dell, and Hewlett Packard. Although any personal computer can run Web server software, it is best to use a computer that has been optimized for this purpose. To be a Web server, a computer must have the Web server software described above installed and be connected to the Internet. Every Web server machine has an IP address. For example, if you type *http://www.aw.com/laudon*, in your browser, the browser software sends a request for HTTP service to the Web server whose domain name is *aw.com*. The server then locates the page named "laudon" on its hard drive, sends the page back to your browser, and displays it on your screen.

Aside from the generic Web server software packages, there are actually many types of specialized servers on the Web, from **database servers** that access specific information with a database, to **ad servers** that deliver targeted banner ads, to **mail servers** that provide mail messages, and **video servers** that provide video clips. At a small e-commerce site, all of these software packages might be running on a single machine, with a single processor. At a large corporate site, there may be hundreds of discrete machines, many with multiple processors, running specialized Web server functions described above.

A **Web client**, on the other hand, is any computing device attached to the Internet that is capable of making HTTP requests and displaying HTML pages. The most common client is a Windows PC or Macintosh, with various flavors of UNIX machines a distant third. However, the fastest growing category of Web clients are not computers at all, but personal digital assistants (PDAs) such as the Palm and HP Jornada, and cellular phones outfitted with wireless Web access software. In general, Web clients can be any device—including a refrigerator, stove, home lighting system, or automobile instrument panel—capable of sending and receiving information from Web servers.

Abbreviations

SSL (Security Socket Layer) 加密套接字协议层   FTP (File Transfer Protocol) 文件传输协议
PDA (personal digital assistant) 个人数字助理

9.4　Web Browsers

The primary purpose of Web browsers is to display Web pages, but browsers also have added features, such as e-mail and newsgroups (an online discussion group or forum).

Currently 94% of Web users use either Internet Explorer or Netscape Navigator, but recently some new browsers have been developed that are beginning to attract attention. The browser Opera is becoming very popular because of its speed—it is currently the world's fastest browser—and because it is much smaller than existing browsers (it can almost fit on a single diskette). It can also remember the last Web page you visited, so the next time you surf, you can start where you left off. And like the big two, you can get it for free; the catch is that you have to watch blinking ads in one corner, or pay $40 for the ad-free version of Opera.

The browser NeoPlanet is also gaining new fans, primarily because of the 500+ *skins*, or design schemes, that come with it. Using skins, you can design the browser to look and sound just the way you'd like it to, rather than being limited to the standard look provided by Navigator and Internet Explorer. However, NeoPlanet requires Internet Explorer's technology in order to operate, so you must also have IE installed on your computer.

New Words & Expressions

extension　　n. 扩展名　　　　　　　supplant　vt. 排挤掉，代替
browser　　　n. 浏览器　　　　　　　surf　　　vi. 作冲浪运动，vt. 在…冲浪
Netscape　　美国 Netscape 公司，以开发 Internet 浏览器闻名

Reading Material: Features of The Internet and The Web

The Internet and the Web have spawned a number of powerful new software applications upon which the foundations of e-commerce are built.

E-MAIL

Since its earliest days, **electronic mail**, or e-mail, has been the most-used application of the Internet. An estimated 3.5 billion business e-mails and 2.7 billion personal emails are sent every day in the United States. Worldwide, more than 8 billion e-mails are sent each day. E-mail uses a series of protocols to enable messages containing text, images, sound, and video clips to be transferred from one Internet user to another. Because of its flexibility and speed, it is now the most popular form of business communication—more popular than the phone, fax, or snail mail (the U.S. Postal Service).

In addition to text typed within the message, e-mail also allows **attachments**, which are files inserted within the e-mail message. The files can be documents, images, or sound or video clips.

Although e-mail was designed to be used for interpersonal messages, it can also be a very effective marketing tool. E-commerce sites purchase e-mail lists from list providers and send mail to prospective customers, as well as existing customers. The response rate from targeted e-mail campaigns can be as high as 20%, extraordinary when compared to banner ad response rates of less than 1%. Most e-commerce sites also have a "Contact Us" section that includes an e-mail contact, to make requests and comments easier for customers.

However, in addition to this acceptable practice of communicating with people who have requested such contact, some companies also use e-mail as a mass mailing technique, also known as **spam,** or unsolicited e-mail. There are a number of state laws against spamming, but it is still the bane of the Web.

SEARCH ENGINES

Search engines can be Web sites themselves, such as Google and AltaVista, or a service within a site that allows users to ask for information about various topics. A **search engine** identifies Web pages that appear to match keywords, also called *queries*, typed by the user and provides a list of the best matches. A query can be a question, a series of words, or a single word for the search engine to look for.

How exactly individual search engines work is a proprietary secret, and at times defies explanation. Some search engines—among them AltaVista—seek to visit every Web page in existence, read the contents of the home page, identify the most common words or keywords, and create a huge database of domain names with keywords. Sometimes the search engines will just read the meta tags and other keyword sections of the home page. This is faster, but Web designers often stuff an extraordinary number of keywords into their meta tags. The program that search engines unleash on the Web to perform this indexing function is called a *spider* or *crawler*. Unfortunately, as the number of Web pages climbs to over two billion, more and more pages are missed by the search engines. Google, perhaps the most complete search engine, contains references to only about half (one billion) of all Web pages. And the engines do not always overlap, which means you may miss a page on one engine, but pick it up on another. It's best therefore to use multiple search engines.

Other search engines use different strategies. Google uses a collaborative filtering technique: It indexes and ranks sites based on the number of users who request and land at a site. This method is biased by volume: You see the Web pages others have asked to see. Yahoo, on the other hand, uses a staff of human indexers to organize as many pages as they can. It is very difficult to get your site registered on Yahoo because of the limitations of their method, which is biased toward sites that somehow come to the attention of Yahoo staff. Once again, the best advice is to use several different search engines.

One of the newest trends in search engines is focus; instead of trying to cover every possible information need that users have, some search engines are electing to specialize in one particular area. By limiting their coverage to such topics as sports, news, medicine, or finance, niche search engines are hoping to differentiate themselves from the crowd and provide better quality results for users. FindLaw.com, a search engine and directory of legal information, has seen its searches rising steadily. The same is true of Moreover.com, a search engine that specializes in collecting and reporting news headlines from more than 1,800 news sites.

Although the major search engines are used for tracking down general information of interest to users, such as a site for buying beer-making supplies, or statistics on Internet usage in Barbados, they have also become a crucial tool within e-commerce sites. Customers can more easily search for the exact item they want with the help of a search program; the difference is that within Web sites, the search engine is limited to finding matches from that one site. Sites without search engines are asking visitors to spend lots of time exploring the site—something few people are willing to do—when most sites offer a quick-and-easy way to find what they're looking for.

INTELLIGENT AGENTS (BOTS)

Intelligent agents, or **software robots** (**bots** for short) are software programs that gather and/or filter information on a specific topic, and then provide a list of results for the user. Intelligent agents were originally invented by computer scientists interested in the development of artificial intelligence (a family of related technologies that attempt to imbue computers with human-like intelligence). However, with the advent of e-commerce on the Web, interest quickly turned to exploiting intelligent agent technology for commercial purposes. Today, there are a number of different types of bots used in e-commerce on the Web, and more are being developed every day.

For instance, as previously noted, many search engines employ *web crawlers* or *spiders* that crawl from server to server, compiling lists of URLs that form the database for the search engine. These web crawlers and spiders are actually bots.

The *shopping bot* is another common type of bot. Shopping bots search online retail sites all over the Web and then report back on the availability and pricing of a range of products. For instance, you can use MySimon.com's shopping bot to search for a Sony digital camera. The bot provides a list of online retailers that carry a particular camera model, as well as report about whether it is in inventory and what the price and shipping charges are.

Another type of bot, called an *update bot*, allows you to monitor for updated materials on the Web, and will e-mail you when a selected site has new or changed information.

News bots will create custom newspapers or clip articles for you in newspapers around the world.

INSTANT MESSAGING

E-mail messages have a time lag of several seconds to minutes between when messages are sent and received, but **instant messaging (IM)** displays words typed on a computer almost instantaneously. Recipients can then respond immediately to the sender the same way, making the communication more like a live conversation than is possible through e-mail.

America Online (AOL) was the first to introduce a widely accepted Instant Messaging system several years ago, which is credited with the company's sudden surge in users. AOL's system is proprietary. One of the key components of an IM service is a *buddy list*, as AOL called it. The buddy list is a private list of people with whom you might want to communicate. If a person is on your buddy list, AOL will alert you when that individual signs on, enabling an IM to be sent.

The downside is that IM systems are proprietary—no standard has been set yet—so that competing sites have created their own IM services. Yahoo has IM, as does MSN, but neither works in conjunction with the others.

Interestingly, despite the wild popularity of such services, no one seems to know yet how to make money from it. AOL, Yahoo, and MSN have all offered IM free to their users and have no immediate plans to start charging a fee. True, it is a marketing draw that brings in new users, but that doesn't necessarily translate into profits.

Nevertheless, some companies have added IM to their Web sites as a means of offering instant access to customer service. For example, Sotheby's, an auction house, encourages visitors to chat live with a Sotheby's representative online. The hope is that by encouraging consumers' need for immediate gratification—whether in the form of a question answered or product ordered—IM will boost revenues and customer satisfaction.

CHAT

Like IM, **chat** enables users to communicate via computer in real time, that is, simultaneously. However, unlike IM, which can only work between two people, chat can occur between several users.

For many Web sites, developing a community of like-minded users has been critical for their growth and success. Just look at eBay.com, which would probably have been unsuccessful without its corps of auction fans, or About.com, which exists to serve communities of consumers with similar interests. Once those community members come together on a site, chat can be a service that enables them to further bond and network, endearing them further to the Web site.

Chat is also used frequently in distance learning, for class discussions and online discussions sponsored by a company. When a celebrity appears on an entertainment Web site, for example, they use chat software in order to see and respond to questions from audience members out in cyberspace.

MUSIC, VIDEO, AND OTHER STANDARD FILES

Although the low bandwidth of Internet I era connections has made audio and video files more difficult to share, with Internet II, these files will become more commonplace. Today it is possible to send and receive files containing music or other audio information, video clips, animation, photographs, and other images, although the download times can be very long, especially for those using a 56 Kbps modem.

Video clips, Flash animations, and photo images are now routinely displayed either as part of Web sites, or sent as attached files. Companies that want to demonstrate use of their product have found video clips to be extremely effective. And audio reports and discussions have also become commonplace, either as marketing materials or customer reports. Photos, of course, have become an important element of most Web sites, helping to make site designs more interesting and eye catching, not to mention helping to sell products, just as catalogs do.

STREAMING MEDIA

Streaming media enables music, video, and other large files to be sent to users in chunks so that when received and played, the file comes through uninterrupted. Streamed files must be viewed "live": They cannot be stored on client hard drives. RealAudio and RealVideo are the most widely used streaming tools. Streaming audio and video segments used in Web ads or CNN news stories are perhaps the most frequently used streaming services.

Macromedia's Shockwave is commonly used to stream audio and video for instructional purposes. Macromedia's Flash vector graphics program is the fastest growing streaming audio and video tool. Flash has the advantage of being built into most client browsers; no plug-in is required to play Flash files.

COOKIES

Cookies are a tool used by Web sites to store information about a user. When a visitor enters a Web site, the site sends a small text file (the cookie) to the user's computer so that information from the site can be loaded more quickly on future visits. The cookie can contain any information desired by the site designers, including customer number, pages visited, products examined, and other detailed information on the behavior of the consumer at the site. Cookies are useful to consumers because the site will recognize returning patrons and not ask them to register again. Cookies can also help personalize a site by allowing the site to recognize returning customers and make special offers to them based on their past behavior at the site. Cookies can also permit customization and market segmentation—the ability to change the product or the price based on prior consumer information. As we will discuss throughout the book, cookies also can pose a threat to consumer privacy, and at times they are bothersome. Many people clear their cookies at the end of every day. Some disable them entirely.

New Words & Expressions

spawn v. 产生
Snail Mail 由邮递员分发传递的传统信件，指其速度慢
Spam 兜售信息，垃圾邮件
defy vt. 不服从，藐视，使...难于
Search engine 搜索引擎
unleash v. 释放
buddy n. <美口>密友，伙伴
Video clip 视频剪辑
CNN 美国有线新闻网络
fax n. vt. 传真
killer app 招人喜爱的应用程序
virtual reality n. 虚拟现实
Flash 由macromedia公司推出的交互式矢量图和Web动画的标准
Cookie 当你访问某个站点时，随某个HTML网页发送到你的浏览器中的一小段信息

attachment n. 附件，附加装置，配属
unsolicited adj. 未被恳求的，主动提供的
bane n. 毒药，祸害
meta n. 元的
overlap v. 与...交迭
chat v. 聊天；n. 聊天
chunk n. 程序块；组块；字节片
patron n. 赞助人（顾客）
sophisticated adj. 复杂的，久经世故的
connectivity n. 连通性

Abbreviations

MSN (Microsoft Network) 微软提供的网络在线服务

VOIP (Voice Over Internet Protocol) 基于网际协议的声音技术

学术论文的英文写作简介

用英语写学术论文的目的主要有两个，一是参加国际学术会议，在会议上宣讲，促进学术交流；二是在国际学术刊物上发表，使国外同行了解自己的研究成果，同样也是出于学术交流的目的。

一、科技论文的结构

不同的学科或领域、不同的刊物对论文的格式有不同的要求，但各个领域的研究论文在文体和语言特点上都有许多共性。一般来说，一篇完整规范的学术论文由以下各部分构成：

Title（标题）
Abstract（摘要）
Keywords（关键词）
Table of contents（目录）
Nomenclature（术语表）
Body（正文）
- Introduction（引言）
- Method（方法）
- Results（结果）
- Discussion（讨论）
- Conclusion（结论）
Acknowledgement（致谢）
Notes（注释）
References（参考文献）
Appendix（附录）

其中，Title，Abstract，Introduction，Method，Result，Discussion，Conclusion 和 Reference 八项内容是必不可少的，其他内容则根据具体需要而定。在这八项内容中，读者最多的是 Title，Abstract 和 Introduction 部分，读者会根据这些内容来决定是否阅读全文。也就是说，一篇研究论文可以赢得多少读者，在很大程度上取决于 Title、Abstract 和 Introduction 的写作质量，这三部分内容的写作技巧将在以后章节中分别介绍。下面简单介绍一下科技论文的正文、结论和结尾等内容。

二、正文

学术论文的正文一般包括 Method，Result，Discussion 三个部分。这三部分主要描述研究课题的具体内容、方法，研究过程中所使用的设备、仪器、条件，并如实公布有关数据和研究结果等。Conclusion 是对全文内容或有关研究课题进行的总体性讨论。它具有严密的科学性和客观性，反映一个研究课题的价值，同时提出以后的研究方向。

为了帮助说明论据、事实，正文中经常使用各种图表。最常用的是附图（Figure）和表（Table），此外还有图解或简图（Diagram）、曲线图或流程图（Graph）、视图（View）、剖面图（Profile）、图案（Pattern）等。在文中提到时，通常的表达法为：

如图 4 所示　　As (is) shown in Fig.4,
如表 1 所示　　As (is) shown in Tab.1,

二、结论

在正文最后应有结论（Conclusions）或建议（Suggestions）。

关于结论可用如下表达方式：

（1）The following conclusions can be drawn from …
由……可得出如下结论。

（2）It can be concluded that …
可以得出结论……

（3）We may conclude that…
We come to the conclusion that…
我们得出如下结论……

（4）It is generally accepted (believed, held, acknowledged) that…
一般认为……（用于表示肯定的结论）

（5）We think (consider, believe, feel) that…
我们认为……（用于表示留有商量余地的结论）

关于建议可用如下表达方式

（1）It is advantageous to (do)

（2）It should be realized (emphasized, stressed, noted, pointed out) that …

（3）It is suggested (proposed, recommended, desirable) that …

（4）It would be better (helpful, advisable) that…

四、结尾部分

1. 致谢

为了对曾给予支持与帮助或关心的人表示感谢，在论文之后，作者通常对有关人员致以简短的谢词，可用如下方式：

I am thankful to sb. for sth

I am grateful to sb. for sth

I am deeply indebted to sb. for sth

I would like to thank sb. for sth

Thanks are due to sb. for sth

The author wishes to express his sincere appreciation to sb. for sth.

The author wishes to acknowledge sb.

The author wishes to express his gratitude for sth.

2. 注释

注释有两种方式，一种为脚注，即将注释放在出现的当页底部；另一种是将全文注释集中在结尾部分。两种注释位置不同，方法一样。注释内容包括：

（1）引文出处。注释方式参见"参考文献"。

（2）对引文的说明，如作者的见解、解释。

（3）文中所提到的人的身份，依次为职称或职务、单位。如：

Professor, Dean of Dept.... University（教授，……大学……系主任）

Chairman, … Company, USA（美国……公司董事长）

（4）本论文是否曾发表过。

3. 参考文献

在论文的最后应将写论文所参考过的主要论著列出，目的是表示对别人成果的尊重或表示本论文的科学根据，同时也便于读者查阅。参考文献的列法如下：

（1）如果是书籍，应依次写出作者、书名、出版社名称、出版年代、页数。如：

Dailey, C.L. and Wood, F.C., **Computation curves for compressible Fluid Problems**, *John Wiley & Sons, Inc. New York,* 1949, pp.37-39

（2）如果是论文，应依次写出作者、论文题目、杂志名称、卷次、期次、页数。如：

Marrish Joseph G.,**Turbulence Modeling for Computational Aerodynamics**, *AIAA J.Vol-21,No.7*, 1983, PP.941-955

（3）如果是会议的会刊或论文集，则应指出会议举行的时间、地点。如：

Proceedings of the Sixth International Conference on Fracture Dec.4-10,1984, New Delhi, India

（4）如果作者不止一人，可列出第一作者，其后加上 et al。如：

Wagner, R.S. et al, ….

（5）在印刷上，论文或著作名称用黑体字，出版社、杂志名称用斜体字。

Exercises

I. Write a summary about how World Wide Web works

II. Please describe the features of the Internet and Web

III. Fill in the blanks in each of the following

1. The primary purpose of Web _____ is to display Web pages.
2. Cookiesare a tool used by Web sites to store information about a _____.
3. Hypertext is a way of formatting_____with embedded_____that connect documents to one another, and that also link pages to other_____such as sound, video, or animation files.
4. _____is the first set of letters at the start of every Web address, followed by the.

Chapter 10　Network Security

学习指导

　　计算机网络是否安全并不是一个绝对的概念，而是取决于所定义的安全方针。事实上，确定网络安全方针很难，企业常常在安全性与易用性之间折衷。常用的安全技术包括：使用密码控制访问、加密与授权、包过滤与防火墙。通过本章学习，读者应该：
- 熟悉安全网络的定义和安全方针的制定等知识与术语；
- 掌握访问控制、加密与授权、包过滤、防火墙技术等知识与术语；
- 掌握科技论文标题的写法。

10.1　Secure Networks and Policies

　　What is a secure network? Can an Internet be made secure? Although the concept of a secure network is appealing to most users, networks cannot be classified simply as secure or not secure because the term is not absolute—each group defines the level of access that is permitted or denied. For example, some organizations store data that is valuable. Such organizations define a secure network to be a system prevents outsiders from accessing the organization's computers. Other organizations need to make information available to outsiders, but prohibit outsiders from changing the data. Such organizations may define a secure network as one that allows arbitrary access to data, but includes mechanisms that prevent unauthorized changes. Still other groups focus on keeping communication private; they define a secure network as one in which no one other than the intended recipient can intercept and read a message. Finally, many large organizations need a complex definition of security that allows access to selected data or services the organization chooses to make public, while preventing access or modification of sensitive data and services that are kept private.

　　Because no absolute definition of secure network exists, the first step an organization must take to achieve a secure system is to define the organization's security policy. The policy does not specify how to achieve protection. Instead, it states clearly and unambiguously the items that are to be protected.

　　Devising a network security policy can be complex because a rational policy requires an organization to assess the value of information. The policy must apply to information stored in computers as well as to information traversing a network.

10.2 Aspects of Security

Defining a security policy is also complicated because each organization must decide which aspects of protection are most important, and often must compromise between security and ease of use. For example, an organization can consider:
- Data Integrity. Integrity refers to protection from change: is the data that arrives at a receiver exactly the same as the data that was sent?
- Data Availability. Availability refers to protection against disruption of service: does data remain accessible for legitimate uses?
- Data Confidentiality and Privacy. Confidentiality and privacy refer to protection against snooping or wiretapping: is data protected against unauthorized access?

10.3 Responsibility and Control

Many organizations discover that they cannot design a security policy because the organization has not specified how responsibility for information is assigned or controlled. The issue has several aspects to consider:
- Accountability. Accountability refers to how an audit trail is kept: which group is responsible for each item of data? How does the group keep records of access and change?
- Authorization. Authorization refers to responsibility for each item of information and how such responsibility is delegated to other: who is responsible for where information resides and how does a responsible person approve access and change?

The critical issue underlying both accountability and authorization is control – an organization must control access to information analogous to the way the organization controls access to physical resources such as offices, equipment, and supplies.

10.4 Integrity Mechanism

Checksums and cyclic redundancy checks (CRC) techniques can be used to ensure the integrity of data against accidental damage. To use such techniques, a sender computes a small, integer value as a function of the data in a packet. The receiver recomputes the function from the data that arrives, and compares the result to the value that the sender computed.

A checksum or CRC cannot absolutely guarantee data integrity for two reasons. First, if malfunctioning hardware changes the value of a checksum as well as the value of the data, it is possible for the altered checksum to be valid for the altered data. Second, if data changes result from a planned attack, the attacker can create a valid checksum for the altered data.

New Words & Expressions:

archive　vt. 存档；n. 档案文件　　　　　　　incur　v. 招致

liability   n. 责任，义务
unauthorized   a. 未被授权的，未经认可的
data availability   数据有效性
accountability   n. 责任，可计算性
authorization   n. 授权，特许
snoop   vi. 探听，调查，偷窃
cyclic redundancy checks   循环码校验
appeal to   v. 呼吁，要求，诉诸，上诉，有吸引力
prevent from   v. 阻止，妨碍
be separated from   和……分离开，和……分散
disruption   n. 中断，分裂，瓦解，破坏
be analogous to   v. 类似于……，与……相似
result from   v. 由…产生

traverse   n. 横贯，横断；vt. 横过，穿过，经过
data integrity   数据完整性
data confidentiality   数据机密性
audit trail   审计追踪，检查跟踪
integrity mechanisms   完整性机制
checksums   n. 检查机，检验和
wiretap   v. n. 搭线窃听，窃听或偷录
focus on   v. 集中
prohibit from   v. 禁止，阻止
arise from   起于，由……出身
be responsible for   v. 对……负责
malfunctioning   n. 故障

Abbreviations:

CRC (Cyclic Redundancy Check)   循环冗余校验法，循环冗余核对

10.5 Access Control and Passwords

Many computer systems use a password mechanism to control access to resources. Each user has a password, which is kept secret. When the user needs to access a protected resource, the user is asked to enter the password.

A simple password scheme works well for a conventional computer system because the system does not reveal the password to others. In a network, however, a simple password mechanism is susceptible to eavesdropping. If a user at one location sends a password across a network to a computer at another location, anyone who wiretaps the network can obtain a copy of the password. Wiretapping is especially easy when packets travel across a LAN because many LAN technologies permit an attached station to capture a copy of all traffic. In such situations, additional steps must be taken to prevent passwords from being reused.

New Words & Expressions：

be susceptible to   易受……的影响
eavesdrop；   vt. vi. 偷听

eavesdropping   n. 窃听；

10.6 Encryption and Privacy

To ensure that the content of a message remains confidential despite wiretapping, the message must be encrypted. In essence, encryption scrambles bits of the message in such a way that only the intended recipient can unscramble them. Someone who intercepts a copy of the encrypted message will not be able to extract information.

Several technologies exist for encryption. In some technologies, a sender and receiver must both have a copy of an encryption key, which is kept secret. The sender uses the key to produce an encrypted message, which is then sent across a network. The receiver uses the key to decode the encrypted message. That is, the encrypt function used by the sender takes two arguments: a key, K, and a message to be encrypted, M. The function produces an encrypted version of the message, E.

$$E=encrypt (K, M)$$

The decrypt function reverses the mapping to produce the original message:

$$M=decrypt (K, E)$$

Mathematically, decrypt is the inverse of encrypt:

$$M=decrypt (K, encrypt (K, M))$$

10.7 Public Encryption

In many encryption schemes, the key must be kept secret to avoid compromising security. One particularly interesting encryption technique assigns each user a pair of keys. One of the user's keys, called the private key, is kept secret, while the other, called the public key, is published along with the name of the user, so everyone knows the value of the key. The encryption function has the mathematical property that a message encrypted with the public key cannot be easily decrypted except with the private key, and a message encrypted with the private key cannot be decrypted except with the public key.

The relationships between encryption and decryption with the two keys can be expressed mathematically. Let M denote a message, pub-u1 denote user 1's public key, and prv-u1 denote user 1's private key. Then

$$M=decrypt (pub-u1, encrypt (prv-u1, M))$$

and

$$M=decrypt (prv-u1, encrypt (pub-u1, M))$$

Revealing a public key is safe because the functions used for encryption and decryption have a one-way property. That is, telling someone the public key does not allow the person to forge a message that appears to be encrypted with the private key.

Public key encryption can be used to guarantee confidentiality. A sender who wishes a message to remain private uses the receiver's public key to encrypt the message. Obtaining a copy of the message as it passes across the network does not enable someone to read the contents because decryption requires the receiver's private key. Thus, the scheme ensures that data remains confidential because only the receiver can decrypt the message.

New Words & Expressions:

encrypt　vt. 加密，将……译成密码

function　n. 函数

confidential　a. 机密的

argument　n. 自变量，用于确定程序或子程序的值

scramble   n. 混乱；vt. 搅乱，使混杂
intercept   vt. 中途拦截；截击；截取
decrypt   vi. 翻译密码（解码）
one-way   a. 单向
extract   vt. 摘录，析取，吸取
unscramble vt. 整理，使回复原状，译出密码
forge   v. 铸造，伪造
mapping   n. 映象，映射
reveal   vt. 揭示，展现，显示
compromise n. 妥协，折衷；v. 妥协，折衷

10.8 Authentication with Digital Signatures

An encryption mechanism can also be used to authenticate the sender of a message. The technique is known as a digital signature. To sign a message, the sender encrypts the message using a key known only to the sender. The recipient uses the inverse function to decrypt the message. The recipient knows who sent the message because only the sender has the key needed to perform the encryption. To ensure that encrypted messages are not copied and resent later, the original message can contain the time and date that the message was created.

Consider how a public key system can be used to provide a digital signature. To sign a message, a user encrypts the message using his or her private key. To verify the signature, the recipient looks up the user's public key and uses it to decrypt the message. Because only the user knows the private key, only the user can encrypt a message that can be decoded with the public key.

Interestingly, two levels of encryption can be used to guarantee that a message is both authentic and private. First, the message is signed by using the sender's private key to encrypt it. Second, the encrypted message is encrypted again using the recipient's public key. Mathematically, double encryption can be expressed as:

$$X = encrypt\ [pub\text{-}u2,\ encrypt\ (prv\text{-}u1,\ M\ )]$$

where M denotes a message to be sent, X denotes the string that results from the double encryption, prv-u1 denotes the sender's private key, and pub-u2 denotes the recipient's public key.

At the receiving end, the decryption process is the reverse of the encryption process. First, the recipient uses his or her private key to decrypt the message. The decryption removes one level of encryption, but leaves the message digitally signed. Second, the recipient uses the sender's public key to decrypt the message again. The process can be expressed as:

$$M = decrypt\ [pub\text{-}u1,\ decrypt\ (prv\text{-}u2,\ X\ )]$$

where X denotes the encrypted string that was transferred across the network, M denotes the original message, prv-u2 denotes the recipient's private key, and pub-u1 denotes the sender's public key.

If a meaningful message results from the double decryption, it must be true that the message was confidential and authentic. The message must have reached its intended recipient because only the intended recipient has the correct private key needed to remove the outer encryption. The message must have been authentic, because only the sender has the private key needed to encrypt the message so the sender's public key will correctly decrypt it.

New Words & Expressions:

authentication   n. 证明（鉴定，辨证）
inverse function   n. 反函数
look up   v. 检查，查找
public key   公开密钥

digital signatures   n. 数字标记图
resend   vt. 再发，再寄
recipient   n. 接收器； a. 能接受的（容纳的）
private key   私有密钥

10.9 Packet Filtering

To prevent each computer on a network from accessing arbitrary computers or services, many sites use a technique known as packet filtering. As Figure 10-1 illustrates, a packet filter is a program that operates in a router. The filter consists of software that can prevent packets from passing through the router on a path from one network to another. A manager must configure the packet filter to specify which packets are permitted to pass through the router and which should be blocked.

A packet filter operates by examining fields in the header of each packet. The filter can be configured to specify which header fields to examine and how to interpret the values. To control which computers on a network can communicate with computers on another, a manager specifies that the filter should examine the source and destination fields in each packet header. In the figure, to prevent a computer with IP address 192.5.48.27 on the right-hand network from communicating with any computers on the left-hand network, a manager specifies that the filter should block all packets with a source address equal to 192.5.48.27. Similarly, to prevent a computer with address 128.10.0.32 on the left-hand network from receiving any packets from the right-hand network, a manager specifies that the filter should block all packets with a destination address equal to 128.10.0.32.

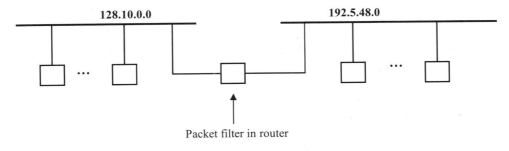

Fig.10-1   A packet filter in a router

In addition to using the source and destination addresses, packet filter can examine the protocol in the packet or the high-level service to which the packet corresponds. The ability to selectively block packets for a particular service means that a manager can prevent traffic to one service, while allowing traffic to another. For example, a manager can configure a packet filter to block all packets that carry World Wide Web communication, while allowing packets that carry e-mail traffic.

A packet filter mechanism allows a manager to specify complex combinations of source and destination addresses, and service. Typically, the packet filter software permits a manager to specify

Boolean combinations of source, destination, and service type. Thus, a manager can control access to specific services on specific computers. For example, a manager might choose to block all traffic destined for the FTP service on computer 128.10.2.14, all World Wide Web traffic leaving computer 192.5.48.33, and all e-mail from computer 192.5.48.34. The filter blocks only the specified combinations—the filter passes traffic destined for other computers and traffic for other services on the specified computers.

New Words & Expressions:

packet filtering　n. 包滤波　　　　　packet filter　n.包滤波器
router　n. 路由器　　　　　　　　　rignt-hand　a. 右手的，得力的
left-hand　a. 左手的，左侧的　　　　packet header　n. 包标题
Boolean combination　n. 布尔组合　　block　n. 块，批；vt. 防碍，阻塞

10.10　Internet Firewall Concept

A packet filter is often used to protect an organization's computers and networks from unwanted Internet traffic. As Figure 10-2 illustrates, the filter is placed in the router that connects the organization to the rest of the Internet.

A packet filter configured to protect an organization against traffic from the rest of the Internet is called an Internet firewall; the term is derived from the fireproof physical boundary placed between two structures to prevent fire from moving between them. Like a conventional firewall, an Internet firewall is designed to keep problems in the Internet from spreading to an organization's computers.

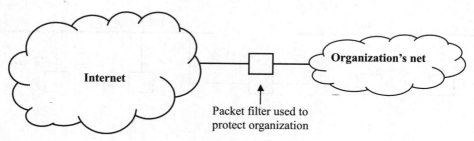

Fig.10-2　A packet filter used as the primary part of a firewall

Firewalls are the most important security tool used to handle network connections between two organizations that do not trust each other. By placing a firewall on each external network connection, an organization can define a secure perimeter that prevents outsiders from interfering with the organization's computers. In particular, by limiting access to a small set of computers, a firewall can prevent outsiders from probing all computers in an organization or flooding the organization's network with unwanted traffic.

A firewall can lower the cost of providing security. Without a firewall to prevent access, outsiders can send packets to arbitrary computers in an organization. Consequently, to provide security, an organization must make all of its computers secure. With a firewall, however, a manager can restrict incoming packets to a small set of computers. In the extreme case, the set can contain a single computer. Although computers in the set must be secure, other computers in the organization do not need to be. Thus, an organization can save money because it is less expensive to install a firewall than to make all computer systems secure.

New Words & Expressions:

Internet Firewall   n. Internet 防火墙   protect somebody from sth. 防止

perimeter   n. 周长（圆度，视野），周界

Reading Material: KINDS OF SECURITY BREACHES

In security, an exposure is a form of possible loss or harm in a computing system; examples of exposures are unauthorized disclosure of data, modification of data, or denial of legitimate access to computing. A vulnerability is a weakness in the security system that might be exploited to cause loss or harm. A human who exploits a vulnerability perpetrates an attack on the system. Threats to computing systems are circumstances that have the potential to cause loss or harm; human attacks are examples of threats, as are natural disasters, inadvertent human errors, and internal hardware or software flaws. Finally, a control is a protective measure-an action, a device, a procedure, or a technique-that reduces a vulnerability.

The major assets of computing systems are hardware, software, and data. There are four kinds of threats to the security of a computing system: interruption, interception, modification, and fabrication. The four threats all exploit vulnerabilities of the assets in computing systems. These four threats are shown in Figure 10-3.

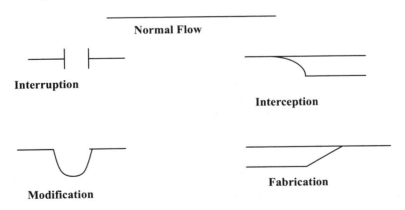

Fig.10-3   Four classes of System Security Failures

(1) In an interruption, an asset of the system becomes lost or unavailable or unusable. An example is malicious destruction of a hardware device, erasure of a program or data file, or failure of

an operating system file manager so that it cannot find a particular disk file.

(2) An interception means that some unauthorized party has gained access to an asset. The outside party can be a person, a program, or a computing system. Examples of this type of failure are illicit copying of program or data files, or wiretapping to obtain data in a network. While a loss may be discovered fairly quickly, a silent interceptor may leave no traces by which the interception can be readily detected.

(3) If an unauthorized party not only accesses but tampers with an asset, the failure becomes a modification. For example, someone might modify the values in a database, alter a program so that it performs an additional computation, or modify data being transmitted electronically. It is even possible for hardware to be modified. Some cases of modification can be detected with simple measures, while other more subtle changes may be almost impossible to detect.

(4) Finally, an unauthorized party might fabricate counterfeit objects for a computing system. The intruder may wish to add spurious transactions to a network communication system. or add records to an existing data base. Sometimes these additions can be detected as forgeries, but if skillfully done, they are virtually indistinguishable from the real thing.

These four classes of interference with computer activity-interruption, interception, modification, and fabrication-can describe the kinds of exposures possible[1].

New Words & Expressions

vulnerability　n. 弱点，攻击
Principle of Easiest Penetration　最易攻破原理
threat　n. 威胁，恐吓
interruption　n. 中断，打断
modification　n. 修改
tamper　v. 篡改

penetrate　vt. 攻破，攻击
exposure　n. 暴露，曝光，揭露
asset　n. 资产，有用的东西
interception　n. 截取
fabricate　v. 伪造
spurious　adj. 假的

Notes

1. possible：后置语修饰 exposures。全句译为：这四种对计算机工作的干扰——中断、截取、修改或伪造——表明了可能出现的几种威胁类型。

科技论文标题的写法

论文标题是全文内容的缩影。读者通过标题便能够预测论文的主要内容和作者的意图，从而决定是否阅读全文。因此，为了使文章赢得有关领域里众多的读者，论文的标题必须用最精炼的语言恰如其分地体现全文的主题和核心。

学术文章的标题主要有三种结构：名词性词组（包括动名词），介词词组，名词词组+介词词组。间或也用一个疑问句作标题（多用在人文社会科学领域），但一般不用陈述句或动词词组作标题。

1. 名词性词组

名词性词组由名词及其修饰语构成。名词的修饰语可以是形容词、介词短语，有时也可以是另一个名词。名词修饰名词时，往往可以缩短标题的长度。以下各标题分别由两个名词词组构成。例如：

Latent demand and the browsing shopper（名词词组+名词词组）

Cost and productivity（名词+名词）

2. 介词词组

介词词组由介词十名词或名词词组构成。如果整个标题就是一个介词词组的话，一般这个介词是"on"，意思是"对……的研究"。例如：

From Knowledge Engineering to Knowledge Management（介词词组+介词词组）

On the correlation between working memory capacity and performance on intelligence tests

3. 名词/名词词组+介词词组

这是标题中用得最多的结构。例如：

Simulation of Controlled Financial Statements（名词+介词词组）

The impact of internal marketing activities on external marketing outcomes（名词+介词词组+介词词组）

Diversity in the Future Work Force（名词+介词词组）

Models of Sustaining Human and Natural Development（名词+介词词组）

标题中的介词词组一般用来修饰名词或名词词组，从而限定某研究课题的范围。这种结构与中文的"的"字结构相似，区别是中文标题中修饰语在前，中心词在后。英文正好相反，名词在前，而作为修饰语的介词短语在后。例如：

Progress on Fuel Cell and its Materials（燃料电池及其材料进展）

4. 其他形式

对于值得争议的问题，偶尔可用疑问句作为论文的标题，以点明整个论文讨论的焦点。例如：

Is B2B e-commerce ready for prime time?

Can ERP Meet Your eBusiness Needs?

有的标题由两部分组成，用冒号（：）隔开。一般来说，冒号前面一部分是研究的对象、内容或课题，比较笼统，冒号后面具体说明研究重点或研究方法。这种结构可再分为三种模式。

模式 1 研究课题：具体内容。例如：

Microelectronic Assembly and Packaging Technology: Barriers and Needs

The Computer Dictionary Project: an update

模式 2 研究课题：方法/性质。例如：

B2B E-Commerce: A Quick Introduction

The Use of Technology in Higher Education Programs: a National Survey

模式 3 研究课题：问题焦点。例如：

Caring about connections: gender and computing

Exercises

I. Answer the following questions

1. What is a secure network?
2. How to ensure the content of a message confidential despite wiretapping?
3. What is a digital signature?
4. What is an Internet firewall and what are its functions?

Chapter 11　Database System

学习指导

数据库是指专门组织起来的一套数据或信息，其目的是为了便于计算机快速查询与检索。一个数据库可以分解为若干个记录，每一个记录又包含若干字段（域）。数据库有多种结构，如文件型、关系型、层次型、网络型与面向对象型等。通过本章学习，读者应掌握：
- 数据库的基本概念与术语；
- 数据库的模型及特点；
- 了解数据挖掘方法；
- 掌握英文论文引言的写作技巧。

11.1　Overview

Database (sometimes spelled data base) is also called an electronic database, referring to any collections of data, or information, that is specially organized for rapid search and retrieval by a computer. Databases are structured to facilitate the storage, retrieval, modification, and deletion of data in conjunction with various data-processing operations. Databases can be stored on magnetic disk or tape, optical disk, or some other secondary storage device.

A database consists of a file or a set of files. The information in these files may be broken down into records, each of which consists of one or more fields. Fields are the basic units of data storage, and each field typically contains information pertaining to one aspect or attribute of the entity described by the database. Using keywords and various sorting commands, users can rapidly search, rearrange, group, and select the fields in many records to retrieve or create reports on particular aggregates of data.

Database records and files must be organized to allow retrieval of the information. Early systems were arranged sequentially (i.e., alphabetically, numerically, or chronologically); the development of direct-access storage devices made possible random access to data via indexes. Queries are the main way users retrieve database information. Typically, the user provides a string of characters, and the computer searches the database for a corresponding sequence and provides the source materials in which those characters appear. A user can request, for example, all records in which the content of the field for a person's last name is the word Smith.

The many users of a large database must be able to manipulate the information within it quickly at any given time. Moreover, large business and other organizations tend to build up many independent files containing related and even overlapping data, and their data processing activities often require the linking of data from several files. Several different types of database management systems have been developed to support these requirements: flat, hierarchical, network, relational, and object-oriented.

In flat databases, records are organized according to a simple list of entities; many simple databases for personal computers are flat in structure. The records in hierarchical databases are organized in a treelike structure, with each level of records branching off into a set of smaller categories. Unlike hierarchical databases, which provide single links between sets of records at different levels, network databases create multiple linkages between sets by placing links, or pointers, to one set of records in another; the speed and versatility of network databases have led to their wide use in business. Relational databases are used where associations among files or records cannot be expressed by links; a simple flat list becomes one table, or "relation", and multiple relations can be mathematically associated to yield desired information. Object-oriented databases store and manipulate more complex data structures, called "objects", which are organized into hierarchical classes that may inherit properties from classes higher in the chain; this database structure is the most flexible and adaptable.

The information in many databases consists of natural-language texts of documents; number-oriented databases primarily contain information such as statistics, tables, financial data, and raw scientific and technical data. Small databases can be maintained on personal-computer systems and may be used by individuals at home. These and larger databases have become increasingly important in business life. Typical commercial applications include airline reservations, production management, medical records in hospitals, and legal records of insurance companies. The largest databases are usually maintained by governmental agencies, business organizations, and universities. These databases may contain texts of such materials as abstracts, reports, legal statutes, wire services, newspapers and journals, encyclopaedias, and catalogs of various kinds. Reference databases contain bibliographies or indexes that serve as guides to the location of information in books, periodicals, and other published literature. Thousands of these publicly accessible databases now exist, covering topics ranging from law, medicine, and engineering to news and current events, games, classified advertisements, and instructional courses. Professionals such as scientists, doctors, lawyers, financial analysts, stockbrokers, and researchers of all types increasingly rely on these databases for quick, selective access to large volumes of information.

New Words & Expressions：

facilitate　vt. 使容易，推动，促进
field　字段，是数据库的基本存取单位
relational database　关系数据库
object-oriented database　面向对象的数据库
chronologically　adv. 按年代顺序排列地
build up　v. 建造，建立，装配，组成

retrieval　n. 检索
record　记录，由字段组成的一种数据单位
flat database　非结构化的数据库
alphabetically　adv. 按字母顺序地
break down　v. 分解
range from…to …　vt. 从…到

encyclopaedia n. 百科全书

11.2 Database Models

There are many different ways to structure and organize the relationships among data in a database. The first step in creating a database is to choose the model with which to represent the data. There are the building blocks that are used to construct more complex database models.

A model, often called a schema, is used to describe the overall characteristics of a database. Like the table of contents for a book, a database model identifies the major parts (e.g., files, records, and fields) of a database and illustrates how these parts fit together.

Database models include flat file, relational, hierarchical, network, object oriented, and text.

11.2.1 Flat File

The simplest of all database models is the flat file, also called a table. The flat file is a single file consisting of rows (records) and columns (fields) of data that resemble a two-dimensional spreadsheet. For example, suppose you want to create a customer file for your mail-order business. You can use a flat-file model with one record per customer, use the names and addresses of your customers to create individual fields, and combine those data with a unique customer identification (ID) field. The ID field is a key field that eliminates the problem of duplicate customer names (e.g., Mark Smith in Seattle and Mark Smith in New York).

Whereas a single flat file such as a customer file is useful for keeping track of customers and preparing mailing lists, it cannot create a complete mail-order application. For that a different database model is needed (as Table 11-1).

11.2.2 Relational

In 1970, E. F. Codd, who was working for IBM at the time, published a paper titled "A Relational Model of Data for Large Shared Data Banks". That paper is now credited as the origin of the entire field of relational database technology.

Table 11-1

Cust. ID	Name	First	Address	City	St	Zip
001	Smith	Mark	656 246th St.	Roslyn	NY	11576
002	Daolt	Shelia	12 Windsong Dr.	Arlington	VA	22201
003	Nasser	James	123 Watercress Ln.	Midvale	UT	84074
004	Smith	Mark	807 W 19th St.	Seattle	WA	98168

The relational model uses one or more flat files or tables and creates relationships among the tables on the basis of a common field in each of the tables. Each flat file or table is called a relation. For example, a mail-order application must organize the customers as well as the products for sale, so what is needed is an inventory file. The inventory file describes each item for sale, and its key field is a unique item number.

The other components of our mail-order application are taking orders and billing customers. For taking orders, an order file would consist of an order number, order data, customer number, item number and quantity ordered. To generate bills or invoices, an invoice file would contain invoices number, customer ID, date, items and quantities ordered, prices, and whether the invoice has been paid.

The database now consists of four flat files: customer, inventory, order, and invoice. Note that there is little redundancy or duplication of data among the relations. Now we are ready for the data manipulation operations.

Filling an order would proceed somewhat as follows. A customer sends in an order for merchandise. That order is assigned a unique order number. The order file is updated by relating the data in the customer file with the data in the inventory file. Similarly, the invoice is generated by relating the inventory file, the order file, and the customer file on the basis of the common fields among the files.

In relational terminology, what has been happening is that a set of operators, known as relational algebra, has been applied to the relations or files. The operators take data in exiting files and produce the desired results. Common operators are the join operator, which combines two separate files using a common field, and the project operator, which creates a new file by selecting fields from an existing file.

Examples of products that use the relational model are IBM's DB2, Oracle Corporation's Oracle, and Sybase, Inc.'s Sybase DBMS .

11.2.3 Hierarchical

The hierarchical model is older than the relational model. It creates relationships among data by structuring data into an inverted tree in which records contain:

(1) A single root or master key field that identifies the type, location, or ordering of the records.

(2) A variable number of subordinate fields that define the rest of the data within a record.

The hierarchical model was developed because hierarchical relationships are commonly found in business applications. As you have known, an organization chart often describes a hierarchical relationship: top management is at the highest level, middle management at lower levels, and operational employees at the lowest levels. Note that within a strict hierarchy, each level of management may have many employees or levels of employees beneath it, but each employee has only one manager. Hierarchical data are characterized by this one-to-many relationship among data.

As another example, consider a simplified airplane spare parts database. An airplane, like most systems, is made up of a set of assemblies that are made up of subassemblies and so on. Using a hierarchical approach, the relationships between records and fields might be established as follows: the first or highest level would contain the major assemblies, such as wings, fuselage, and cockpit; the second level of the hierarchy would contain subassemblies for each major assembly; and the levels farther down would include the specific part numbers and part information.

This approach would be very convenient for answering customer inquiries about parts and their availability, but it would be less convenient for making an inquiry about what parts are on what planes. Before information about specific parts and planes can be obtained, each major assembly must first be retrieved, and several levels of the hierarchy must be navigated to obtain the part information.

In the hierarchical approach, each relationship must be explicitly defined when the database is created. Each record in a hierarchical database can contain only one key field and only one relationship is allowed between any two fields. This can create a problem because data do not always conform to such a strict hierarchy. The rivets in a wing, for example, might be identical to the rivets in the fuselage.

Examples of commercial database products that use the hierarchical model are IBM's IMS and Cullinet's IDMS.

11.2.4 Other Database Models

There are several other database models that are worth a brief mention. We have already mentioned the text model.

The network model creates relationships among data through a linked-list structure in which subordinate records can be linked to more than one parent record. This approach combines records with links, which are called pointers. The pointers are addresses that indicate the location of a record. With the network approach, a subordinate record can be linked to a key record and at the same time itself be a key record linked to other sets of subordinate records. The network model historically has had a performance advantage over other database models. Today, such performance characteristics are only important in high-volume, high-speed transaction processing such as automatic teller machine networks or airline reservation system.

The object-oriented model groups data into collections that represent some kind of object. For example, all the data that relate to the wing of an airplane. More important, the object-oriented model also allows records to inherit information from ancestor records. In the mail-order application, for example, an order for shoes would inherit information such as the customer's name and address from the order file.

New Words & Expressions:

inventory   n. 详细目录，存货，清单
hierarchical   a. 层次的
mail-order   邮购
on the basis of   在……的基础上，基于
subordinate   a. 次要的，从属的，下级的；n. 下属
transaction   n. 办理，处理，交易，处理事务
merchandise   n. 商品，货物
explicitly   ad. 明白地，明确地

invoice   n. 发票，发货单
fuselage   n. 机身
inherit   vt. 继承
master key   万能钥匙，主健
resemble   vt. 象，类似
retrieve   v. 重新得到，检索
ID (identification)   身分，身份
be made up of   由……组成

11.3 Data Mining

A rapidly expanding subject that is closely associated with database technology is data mining, which consists of techniques for discovering patterns in collections of data. Data mining has become an important tool in numerous areas including marketing, inventory management, quality control, loan risk management, fraud detection, and investment analysis. Data mining techniques even have applications in what might seem unlikely settings as exemplified by their use in identifying the functions of particular genes encoded in DNA molecules and characterizing properties of organisms.

Data mining activities differ from traditional database interrogation in that data mining seeks to identify previously unknown patterns as opposed to traditional database inquiries that merely ask for the retrieval of stored facts. Moreover, data mining is practiced on static data collections, called data warehouses, rather than "online" operational databases that are subject to frequent updates. These warehouses are often "snapshots" of databases or collections of databases. They are used in lieu of the actual operational databases because finding patterns in a static system is easier than in a dynamic one.

We should also note that the subject of data mining is not restricted to the domain of computing but has tentacles that extend far into statistics. In fact, many would argue that since data mining had its origins in attempts to perform statistical analysis on large, diverse data collections, it is an application of statistics rather than a field of computer science.

Two common forms of data mining are class description and class discrimination. Class description deals with identifying properties that characterize a given group of data items, whereas class discrimination deals with identifying properties that divide two groups. For example, class description techniques would be used to identify characteristics of people who buy small economical vehicles, whereas class discrimination techniques would be used to find properties that distinguish customers who shop for used cars from those who shop for new ones.

Another form of data mining is cluster analysis, which seeks to discover classes. Note that this differs from class description, which seeks to discover properties of members within classes that are already identified. More precisely, cluster analysis tries to find properties of data items that lead to the discovery of groupings. For example, in analyzing information about people's ages who have viewed a particular motion picture, cluster analysis might find that the customer base breaks down into two age groups—a 4 to10 age group and a 25 to 40 age group. (Perhaps the motion picture attracted children and their parents?) Still another form of data mining is association analysis, which involves looking for links between data groups. It is association analysis that might reveal that customers who buy potato chips also buy beer and soda or that people who shop during the traditional weekday work hours also draw retirement benefits.

Outlier analysis is another form of data mining. It tries to identify data entries that do not comply with the norm. Outlier analysis can be used to identify errors in data collections, to identify credit card theft by detecting sudden deviations from a customer's normal purchase patterns, and perhaps to identify potential terrorists by recognizing unusual behavior.

Finally, the form of data mining called sequential pattern analysis tries to identify patterns of behavior over time. For example, sequential pattern analysis might reveal trends in economic systems such as equity markets or in environmental systems such as climate conditions.

As indicated by our last example, results from data mining can be used to predict future behavior. If an entity possesses the properties that characterize a class, then the entity will probably behave like members of that class. However, many data mining projects are aimed at merely gaining a better understanding of the data, as witnessed by the use of data mining in unraveling the mysteries of DNA. In any case, the scope of data mining applications is potentially enormous, and thus data mining promises to be an active area of research for years to come.

Note that database technology and data mining are close cousins, and thus research in one will have repercussions in the other. Database techniques are used extensively to give data warehouses the capability of presenting data in the form of data cubes (data viewed from multiple perspectives—the term cube is used to conjecture the image of multiple dimensions) that make data mining possible. In turn, as researchers in data mining improve techniques for implementing data cubes, these results will pay dividends in the field of database design.

In closing, we should recognize that successful data mining encompasses much more than the identification of patterns within a collection of data. Intelligent judgment must be applied to determine whether those patterns are significant or merely coincidences. The fact that a particular convenience store has sold a high number of winning lottery tickets should probably not be considered significant to someone planning to buy a lottery ticket, but the discovery that customers who buy snack food also tend to buy frozen dinners might constitute meaningful information to a grocery store manager. Likewise, data mining encompasses a vast number of ethical issues involving the rights of individuals represented in the data warehouse, the accuracy and use of the conclusions drawn, and even the appropriateness of data mining in the first place.

New Words & Expressions

fraud　n. 欺诈，欺骗行为；骗子
DNA　n. 脱氧核糖核酸
snapshots　n. 快照
tentacle　n. 触手，触角，触须
cluster　n. 簇，团，群，组
enormous　adj. 巨大的，极大的，庞大的
conjecture　n. 推测，猜想；vt. & vi. 推测，猜想
encompass　vt. 围绕；包围，包含，包括，涉及

gene　n. 基因
warehouse　n. 仓库，货栈
lieu　n. 代替，（以…）替代
statistics　n. 统计数据，统计学
outlier　n. 局外人，外露层
perspective　n. 远景，透视，观点，想法
appropriateness　n. 得体

英文论文引言的写作技巧

学术论文中的引言（Introduction）是对全文内容和结构的总体勾画。引言尽管不像摘要那

样有一定的篇幅限制和相对固定的格式，但在内容和结构模式上也有需要遵循的规律。

一、引言的内容与结构布局

引言的主要任务是向读者勾勒出全文的基本内容和轮廓。它可以包括以下五项内容中的全部或其中几项：

- 介绍某研究领域的背景、意义、发展状况、目前的水平等；
- 对相关领域的文献进行回顾和综述，包括前人的研究成果，已经解决的问题，并适当加以评价或比较；
- 指出前人尚未解决的问题，留下的技术空白，也可以提出新问题、解决这些新问题的新方法、新思路，从而引出自己研究课题的动机与意义；
- 说明自己研究课题的目的；
- 概括论文的主要内容，或勾勒其大体轮廓。

如何合理安排以上这些内容，将它们有条有理地给读者描绘清楚，并非容易之事。经验告诉我们，引言其实是全文最难写的一部分。这是因为作者对有关学科领域的熟悉程度，作者的知识是渊博、还是贫乏，研究的意义何在、价值如何等问题，都在引言的字里行间得以充分体现。

我们可以将引言的内容分为三到四个层次来安排：

第一层由研究背景、意义、发展状况等内容组成，其中还包括某一研究领域的文献综述；

1) Introducing the general research area including its background, importance, and present level of development.

2) Reviewing previous research in this area.

第二层提出目前尚未解决的问题或急需解决的问题，从而引出自己的研究动机与意义；

Indicating the problem that has not been solved by previous research, raising a relevant question.

第三层说明自己研究的具体目的与内容；

Specifying the purpose of your research.

最后是引言的结尾，可以介绍一下论文的组成部分。

1) Announcing your major findings.

2) Outlining the contents of your paper.

值得注意的是，引言中各个层次所占的篇幅可以有很大差别。这一点与摘要大不一样，摘要中的目的、方法、结果、结论四项内容各自所占的篇幅大体比例一样。而在引言中，第一个层次往往占去大部分篇幅。对研究背景和目前的研究状况进行较为详细的介绍。研究目的可能会比较简短。

引言与摘要还有一点不同的是，摘要中必须把主要研究结果列出，而在引言中（如果摘要与正文一同登出）结果则可以省略不写，这是因为正文中专门有一节写结果（results），不必在引言中重复。

下面这段引言的例子摘自一篇关于混合电动汽车的研究论文，大部分篇幅介绍研究背景。

A Hybrid Internal Combustion Engine/Battery Electric Passenger Car for Petroleum Displacement

I. Forster and J. R. Bumby

INTRODUCTION

[1] The finite nature of the world's oil resources and the general concern about automobile emissions（排放）have prompted the adoption of energy conservation policies and emphasized the need to transfer energy demand from oil to other sources of energy, such as natural gas, coal and nuclear. [2] A transfer of energy from oil to electricity can be achieved to a limited extent in the road transport sector by the increased use of electric vehicle. However, such vehicles are limited in range due to the amount of energy that can he realistically stored on board（在车上）the vehicle without affecting payload（有效载荷）. As a consequence of this, electric vehicles must he used in situations where daily usage is well defined, for example, in urban delivery duty. Indeed, it has been in such vehicles as the urban milk delivery vehicles that electric traction drives have been traditionally applied with a great deal of success. Currently the demand is for urban electric vehicles to he developed with greater traffic compatibility in terms of speed and range.

[3] Although urban delivery vehicle applications will help to reduce the dependence of the road transport sector on petroleum-based fuels, the major part of this market requires vehicles that are not limited in range and have a performance compatible with internal combustion（内燃机）, i.e. engine vehicles. The use of advanced traction battery technology to overcome the range limitation of electric vehicles is one possible solution. However, this would still result in a vehicle limited in range and may in itself create additional problems. For example, due to the much greater on-board stored energy, the charging time required will be greater than at present. [4] The range limitations of the pure electric vehicle can be overcome by using a hybrid i.e. engine/electric drive which incorporates both an i.e. engine and an electric traction system. Although such a vehicle can be designed to meet a number of objectives, it has been argued that a vehicle which seeks to remove the range limitation of the electric vehicle while substituting a substantial amount of petroleum fuel by electrical energy is the vehicle most worth pursuing. With the emphasis of the vehicle design on the electric drive train, the intention may be to operate in an all-electric mode under urban conditions and to use the i.e. engine for long-distance motorway driving. The hybrid mode could then he used for extending urban range and/or improving vehicle accelerative performance on accelerator kick-down.[5] The concept of a hybrid electric vehicle capable of substituting petroleum fuel is' not new, Bosch and Volkswagen（大众汽车）having built vehicles in the 1970s. More recently, the advent of the Electric and Hybrid Vehicle Research, Development and Demonstration Program in the United States of America initiated the design and construction of a Near Term Hybrid Vehicle (NTHV) with the principal aim of substituting petroleum fuel by ' wall plug' electricity. [6] As part of the NTHV program, a large number of conceptual studies were conducted but on vehicles aimed at the American passenger car market. In this paper optimization studies were conducted, but now on a vehicle suitable for the European medium-sized passenger car market. Such optimization studies are important as, with two sources of traction power available, the way in which they are controlled,

and their relative sizing, is fundamental to the way the vehicle performs.[7] Before examining in detail the optimum control strategy for the drive train, Section 2 defines the hybrid arrangement under study. A description of the optimization process using an appropriate cost function is then presented in Section 3 followed by a method of translating the resulting control structure into a sub-optimum algorithm capable of being implemented in real time. Using the optimum control structure the effect of component ratings on the vehicle's performance is evaluated in Section 4, while Section 5 discusses the practical implementation of an overall vehicle control algorithm. Finally, in Section 6, an indication of the vehicle's potential for substituting petroleum fuel by electricity is given.

分析：第一层（第1—5段）：介绍混合电动汽车的研究背景、意义、目前的发展水平，需要解决的问题等。第1段：指出混合电动汽车的研究背景。世界石油资源的有限性及人们对汽车排放问题的广泛关注使得能源转换问题尤为重要。第2段：使用电动汽车能够从某种程度上实现能源转换。但问题是电动汽车的续驶里程比内燃机车短。所以目前要解决的问题是提高电动汽车的速度和连续行驶里程。第3段：市场要求电动汽车的续驶里程及工作性能与内燃机汽车媲美，但是，即使先进电池可以提高电动汽车的续驶里程，也还会有一些问题不能解决。第4段：续驶里程可以通过使用混合电动汽车来提高。混合电动汽车上既装有内燃机，又装有电动驱动系统，在必要时使用其中一种系统。第5段：回顾并评述前人关于混合电动汽车的研究成果。第二层（第6段第1句话）：指出前人研究的范围具有局限性。关于混合电动汽车的研究只针对美国市场。针对欧洲市场的研究还是空白。第三层（In this paper）本文的研究目的及意义：对欧洲中型客车市场进行优化研究。第四层为第7段：本文的组成部分及各部分的主要内容。

比较简短的论文，引言也可以相对比较简短。为了缩短篇幅，可以用一两句话简单介绍一下某研究领域的重要性、意义或需要解决的问题等。接着对文献进行回顾。然后介绍自己的研究动机、目的和主要内容。至于研究方法、研究结果及论文的组成部分则可以完全省略。

二、如何写引言的开头

引言开头（即第一层）最主要目的是告诉读者论文所涉及的研究领域及其意义是什么，研究要解决什么问题，目前状况或水平如何。也就是说，开头要回答如下问题：

What is the subject of the research?

What is the importance of this subject?

How is the research going at present?

In what way is it important, interesting, and worth studying?

What problem does the research solve?

引言的开头常用句型有：

句型1：研究主题+谓语动词 be…

例：Fuel cell（燃料电池）is a technology for the clean and efficient conversion from chemical energy in fossil fuels to electricity.

句型2：研究主题+ has become …

例：Semiconductor based industry（基于半导体的工业）has be come the largest industry for the USA and it has influenced every other industry and every aspect of human life.

句型 3：研究主题+被动语态

例：b. Air pollution has been extensively studied in recent years.

句型 4：Recently, there has been growing interest in / concern about + 研究主题

例：In the 1990s there has been growing interest in the development of electric vehicles in response to the public demand for cleaner air.

句型 5：Recently there have / has been extensive / increasing /numerous publications / literature/ reporting on + 研究领域

例：There has been increasing reporting about forest decline in North America.

句型 6：Researchers have become increasingly interested in +研究领域

或：Researchers have recently focused their attention on +研究领域

Researchers are recently paying more attention to + 研究领域

例：Researchers have become more interested in environmental indicators.

三、如何写文献综述

文献综述是学术论文的重要组成部分，是作者对他人在某研究领域所做的工作和研究成果的总结与评述，包括他人有代表性的观点或理论、发明发现、解决问题的方法等。在援引他人的研究成果时，必须标注出处，即这一研究成果由何人在何时何地公开发表。

1. 文献出处的标注

引用文献时，不同的学科或领域可能采用各自约定俗成的体系或格式。在写论文时，应该了解自己学科采用的固定格式。目前最常见的体系有两种，一种是作者+出版年体系，另一种是顺序编码体系。下面对这两种体系分别加以介绍。

第一种体系的主要框架模式如下：

模式 1：作者（年代）+谓语动词主动语态+研究内容/成果

例：Hanson et al. (1976) noted that oak mortality and decline were associated with drought and insects throughout a multi-state region of the mid-west.

模式 2：研究内容 / 成果+谓语动词被动语态+（作者年代）

例：Success at this Science Day was found to be linked to parental support (Czemiak 1996).

模式 3：It has been +谓语动词被动语态+by 作者（年代）+that 从句

或：It has been+谓语动词被动语态+that 从句（作者年代）

研究内容 / 成果+谓语动词被动语态+by 作者（年代）

例：It was found by Czemiak (1996) that success at this Science Day was linked to parental support.

如果引用的文献有两个以上的作者，只标明第一作者，后面用拉丁文 et al'表示，意思是"等人""其他人"。

如果在综述中涉及几个项目或文献时，则将这些文献并列标注，必要时用逗号隔开。标注参考文献另一种常见体系是按文献出现的先后顺序编号，置于方括号中，标在指引部分的右上角。被引用的作者、文献名、出版时间、地点等列入论文后面的参考文献中。其顺序要与正

文中标注的顺序一致。

2. 文献综述中的动词运用技巧

（1）两类动词。

从例中我们可以发现，文献综述中常用 state，note，observe，discuss，establish，find，present 等动词。这些动词有两种特性，一种是描述性动词，客观地向读者介绍他人的工作；另一种是评价性动词，在一定程度上代表了作者对他人的工作的理解、解释或态度。文献综述中常用的描述性动词有：describe，discuss，explain，examine，present，state 等。常见的评价性动词有：affirm，allege，argue，assume，claim，imply，maintain，presume，reveal，suggest 等。

（2）动词时态。

文献综述中最常见的时态是一般现在时、一般过去时和现在完成时三种时态。使用不同的动词时态会给句子的意义带来变化，基本原则如下：

原则 1：当作者引用某人过去某个时间所做过的某一项具体的研究时，用一般过去时。如：It was found that success at this Science Day was linked to parental support (Czemiak 1996)...

原则 2：在概括或总结某一研究领域里所做过的一些研究时，用现在完成时。

原则 3：在谈及目前的知识水平、技术水平或存在的问题时，用一般现在时态。

四、如何写研究动机与目的

在介绍了他人在某领域的工作和成果之后，下一步便介绍作者自己的研究动机、目的与内容。介绍研究动机可以从两个角度入手，一是指出前人尚未解决的问题或知识的空白，二是说明解决这一问题，或填补知识空白的重要意义。

主要句型有：

句型 1：用表示否定意义的词例 little，few，no 或 none of+名词作主语，表示"在特定的范围中还没有……"的意思，由此来暗示知识的空白部分。如：

However, there exists little research on science fair projects.

句型 2：用表示对照的句型。如：

a. The research has tended to focus on…, rather than on…

b. These studies have emphasized…, as opposed to…

c. Although considerable research has been devoted to…, rather less attention has been paid to…

d. Although there is much hope that three-dimensional coupled models will lead to better understanding of the factors that control hurricane intensity and to increased reliability of hurricane intensity forecasts, the present generation of models may not have enough horizontal resolution to capture the full intensity of extreme storms.

句型 3：提出问题或假设。如：

a. However, it remains unclear whether…

b. It would thus be of interest to learn how…

c. If these results could be confirmed, they would provide strong evidence for …

d. These findings suggest that this treatment might not be so effective when applied to…

e. It would seem, therefore, that further investigations are needed in order to…

指出或暗示了知识领域里的空白，或提出了问题或假设之后，下一步理所当然应该告诉读者本研究的目的和内容，要解决哪些问题，以填补上述空白，或者证明所提出的假设。如何写摘要中的目的部分。这里只略举几例：

a. The aim of the present paper is to give …
b. This paper reports on the results obtained…
c. In this paper we give preliminary results for…
d. The main purpose of the experiment reported here was to…
e. This study was designed to evaluate…
f. The present work extends the use of the last model by…
g. We now report the interaction between…
h. The primary focus of this paper is on…
i. The aim of this investigation was to test…
j. It is the purpose of the present paper to provide…

从暗示知识的空白到本研究的目的与内容一般需要用一些过度词，以提示一下读者。如上面这些例句中，用了 this paper，here，the present work，now，this investigation，the present paper 等词或词组。为了引起读者的注意，这些词或词组一般放在句首。

如果一项研究、一篇论文不止一个目的，应该按目的的主次排列顺序，并用连接词或词组。常见的连接词还有 additionally，in addition to this，besides this，also，not only…but also…，further，furthermore，moreover 等

五、如何写引言的结尾

研究目的完全可以作为引言的结尾。也可以简单介绍一下文章的结构及每一部分的主要内容，从而起到画龙点睛的作用，使读者了解文章的轮廓和脉络。

至于研究结果，在引言中完全可以不写。研究结果是结论部分最主要的组成部分。下面的例子是引言的结尾，介绍文章的结构。

…Before examining in detail the optimum control strategy for the drive train, Section 2 defines the hybrid arrangement under study. A description of the optimization process using an appropriate cost function is then presented in Section 3 followed by a method of translating the resulting control structure into a sub-optimum algorithm capable of being implemented in real time. Using the optimum control structure the effect of component ratings on the vehicle's performance is evaluated in Section 4, while Section 5 discusses the practical implementation of an overall vehicle control algorithm. Finally, in Section 6, an indication of the vehicle's potential for substituting petroleum fuel by electricity is given.

在介绍全文的结构时，要避免使用同一个句型结构，如，Sections 1 describes…Section 2 analyses…Section 3 discusses…Section 4 summarizes…这样，每句话用同样的词开头，句型结构显得单调、枯燥乏味。

Exercises

Answer the following questions

1. What is a Database?
2. How many types of Database models are used?
3. Why is data mining not conducted on "online" databases?
4. Give an additional example of a pattern that might be found by each of the types of data mining identified in the text.
5. How does data mining differ from traditional database inquiries?

Chapter 12 Multimedia and Compuer Animations

学习指导

多媒体是集文本、声音、图片、动画以及图像于一体的信息显示，它包括视元、声元和组元等要素。通过本章的学习，读者应掌握：
- 多媒体的概念及其要素等知识与术语；
- 各种图象文件格式的含义；
- 计算机动画的设计顺序和功能；
- 英文摘要的写作技巧。

12.1 Multimedia

Multimedia is the presentation of information using the combination of text, sound, pictures, animation, and video. Common multimedia computer applications include games, learning software, and reference materials. Most multimedia applications include predefined associations, known as hyperlinks, that enable users to switch between media elements and topics.

Thoughtfully presented multimedia can enhance the scope of presentation in ways that are similar to the roving associations made by the human mind. Connectivity provided by hyperlinks transforms multimedia from static presentations with pictures and sound into an endlessly varying and informative interactive experience.

Multimedia applications are computer programs; typically they are stored on compact discs (CD-ROMs). They may also reside on the World Wide Web, which is the media-rich component of the international communication network known as the Internet. Multimedia documents found on the World Wide Web are called Web Pages. Linking information together with hyperlinks is accomplished by special computer programs or computer languages. The computer language used to create Web pages is called Hyper-text Markup Language (HTML).

Multimedia applications usually require more computer memory and processing power than the same information represented by text alone. For instance, a computer running multimedia applications must have a fast central processing unit (CPU), which is the electronic circuitry that provides the computational ability and control of the computer. A multimedia computer also requires extra electronic memory to help the CPU in making calculations and to enable the video screen to draw complex images. The computer also needs a high capacity hard disk to store and retrieve multimedia information, and a compact disk drive to play CDROM applications. Finally, a multimedia computer must have a keyboard and a pointing device, such as a mouse or a trackball, so that the user can direct the associations between multimedia elements.

12.1.1 Visual Elements

The larger, sharper, and more colorful an image is, the harder it is to present and manipulate on a computer screen. Photographs, drawings, and other still images must be changed into a format that the computer can manipulate and display. Such formats include bit-mapped graphics and vector graphics.

Bit-mapped graphics store, manipulate, and represent images as rows and columns of tiny dots. In a bit-mapped graphic, each dot has a precise location described by its row and column, much like each house in a city has a precise address. Some of the most common bit-mapped graphics formats are called Graphical Interchange Format (GIF), Tagged Image File Format (TIFF), and Windows Bitmap (BMP).

Vector graphics use mathematical formulas to recreate the original image. In a vector graphic, the dots are not defined by a row-and-column address; rather they are defined by their spatial relationships to one another. Because their dot components are not restricted to a particular row and column, vector graphics can reproduce images more easily, and they generally look better on most video screens and printers. Common vector graphics formats are Encapsulated Postscript (EPS), Windows Metafile Format (WMF), Hewlett-Packard Graphics Language (HPGL), and Macintosh graphics file format (PICT).

Obtaining, formatting, and editing video elements require special computer components and programs. Video files can be quite large, so they are usually reduced in size using compression, a technique that identifies a recurring set of information, such as one hundred black dots in a row, and replaces it with a single piece of information to save space in the computer's storage systems.

Common video compression formats are Audio Video Interleave (AVI), Quicktime, and Motion Picture Experts Group (MPEG). These formats can shrink video files by as much as 95 percent, but they introduce varying degrees of fuzziness in the images.

Animation can also be included in multimedia applications to add motion to images. Animations are particularly useful to simulate real-world situations, such as the flight of a jet airplane. Animation can also enhance existing graphics and video elements adding special effects such as morphing, the blending of one image seamlessly into another.

12.1.2 Sound Elements

Sound, like visual elements, must be recorded and formatted so the computer can understand and use it in presentations. Two common types of audio format are Waveform (WAV) and Musical Instrument Digital Interface (MIDI). WAV files store actual sounds, much as music CDs and tapes do. WAV files can be large and may require compression. MIDI files do not store the actual sounds, but rather instructions that enable devices called synthesizers to reproduce the sounds or music. MIDI files are much smaller than WAV files, but the quality of the sound reproduction is not nearly as good.

12.1.3 Organizational Elements

Multimedia elements included in a presentation require a framework that encourages the user to learn and interact with the information. Interactive elements include pop-up menus, small windows that appear on the computer screen with a list of commands or multimedia elements for the user to choose. Scroll bars, usually located on the side of the computer screen, enable the user to move to another portion of a large document or picture.

The integration of the elements of a multimedia presentation is enhanced by hyperlinks. Hyperlinks creatively connect the different elements of a multimedia presentation using colored or underlined text or a small picture, called an icon, on which the user points the cursor and clicks on a mouse. For example, an article on President John F. Kennedy might include a paragraph on his assassination, with a hyperlink on the words the Kennedy funeral. The user clicks on the hyperlinked text and is transferred to a video presentation of the Kennedy funeral. The video is accompanied by a caption with embedded hyperlinks that take the user to a presentation on funeral practices of different cultures, complete with sounds of various burial songs. The songs, in turn, have hyperlinks to a presentation on musical instruments. This chain of hyperlinks may lead users to information they would never have encountered otherwise.

12.1.4 Multimedia Applications

Multimedia has had an enormous impact on education. For example, medical schools use multimedia-simulated operations that enable prospective surgeons to perform operations on a computer-generated "virtual" patient. Similarly, students in engineering schools use interactive multimedia presentations of circuit design to learn the basics of electronics and to immediately implement, test, and manipulate the circuits they designed on the computer. Even in elementary schools, students use simple yet powerful multimedia authoring tools to create multimedia presentation that enhance reports and essays.

Multimedia is also used in commercial applications. For instance, some amusement arcades offer multimedia games that allow players to race Indy cars or battle each other from the cockpits of make-believe giant robots. Architects use multimedia presentations to give clients tours of houses that have yet to be built. Mail-order businesses provide multimedia catalogues that allow prospective buyers to browse virtual showrooms.

New Words & Expressions:

multimedia n. 多媒体	animation n. 动画
hyperlink n. 超链接	roving a. 流动的
media-rich a. 媒体丰富的	bit-mapped a. 位映像的，位图的
tagged a. 加了标记的	postscript n. 附录
interleave n. 交替	seamlessly ad. 无接缝地
morphing n.（形态）拟合	blending n. 合成

synthesizer   n. 混音器
assassination   n. 暗杀
arcade   n. 拱廊，有拱廊的街道
make-believe   a. 假装的
such as   例如……，象这种的
framework   n. 构架，框架，结构
BMP   位图文件的扩展名

scroll bar   n. 滚动条
computer-generated   a. 计算机产生的
cockpit   n. 驾驶员座舱，战场
for instance   比如
as much as   高达…
pop-up   弹出
WAV   声音资源文件

Abbreviations:

GIF (Graphical Interchange Format) 可交换的图像文件，可交换的图像文件格式
TIFF (Tagged Image File Format) 标签图像文件格式
EPS (Encapsulated Postscript) 附录显示格式
WMF (Windows Metafile Format) Windows 图元文件格式。
HPGL (Hewlett-Packard Graphics Language) 惠普图形语言
AVI (Audio Video Interleave) 多媒体文件格式
MIDI (Musical Instrument Data Interface) 乐器数字界面

12.2 Computer Animation

Some typical applications of computer-generated animation are entertainment (motion pictures and cartoons), advertising, scientific and engineering studies, and training and education. Although we tend to think of animation as implying object motions, the term **computer animation** generally refers to any time sequence of visual changes in a scene. In addition to changing object position with translations or rotations, a computer-generated animation could display time variations in object size, color, transparency, or surface texture. Advertising animations often transfer one object shape into another. For example: transforming a can of motor oil into an automobile engine. Computer animations can also be generated by changing camera parameters, such as position, orientation, and focal length. And we can produce computer animations by changing lighting effects or other parameters and procedures associated with illumination and rendering.

12.2.1 Design of Animation Sequences

In general, a sequence is designed with the following steps:
- Storyboard layout
- Object definitions
- Key-frame specifications
- Generation of in-between frames

This standard approach for animated cartoons is applied to other animation applications as well, although there are many special applications that do not follow this sequence. Real-time computer animations produced by flight simulators, for instance, display motion sequences in response to settings on the aircraft controls [1]. And visualization applications are generated by the solutions of the numerical models. For frame-by-frame animations, each frame of the scene is separately generated and stored. Later, the frame can be recorded on film or they can be consecutively displayed in "real-time playback" mode [2].

The *storyboard* is an outline of the action. It defines the motion sequence as a set of basic events that are to take place. Depending on the type of animation to be produced, the storyboard could consist of a set of rough sketches or it could be a list of the basic ideas for the motion.

An *object* definition is given for each participant in the action. Object can be defined in terms of basic shapes, such as polygons or splines. In addition, the associated movements for each object are specified along with the shape.

A *key frame* is a detailed drawing of the scene at a certain time in the animation sequence. Within each key frame, each object is positioned according to the time for that frame. Some key frames are chosen at extreme positions in the action; others are spaced so that the time interval between key frames is not too great. More key frames are specified for intricate motions than for simple, slowly varying motions.

*In-between*s are the intermediate frames between the key frames. The number of in-betweens needed is determined by the media to be used to display the animation. Film requires 24 frames per second, and graphics terminals are refreshed at the rate of 30 to 60 frames per second. Typically, time intervals for the motion are set up so that there are from three to five in-betweens for each pair of key frames. Depending on the speed specified for the motion, some key frames can be duplicated. For a 1-minute film sequence with no duplication, we would need 1440 frames. With five in-betweens for each pair of key frames, we would need 288 key frames. If the motion is not too complicated, we could space the key frames a little farther apart.

There are several other tasks that may be required, depending on the application. They include motion verification, editing, and production and synchronization of a soundtrack. Many of the functions needed to produce general animations are now computer-generated.

12.2.2 General Computer–Animation Functions

Some steps in the development of an animation sequence are well suited to computer solution, These include object manipulations and rendering, camera motions, and the generation of in-betweens. Animation packages, such as Wave-front, for example, provide special functions for designing the animation and processing individual objects.

One function available in animation packages is provided to store and manage the object database. Object shapes and associated parameters are stored and updated in the database. Other object functions include those for motion generation and those for object rendering. Motions can be generated according to specified constraints using two-dimensional or three-dimensional transformations. Standard functions can then be applied to identify visible surfaces and apply the rendering algorithms.

Another typical function simulates camera movements. Standard motions are zooming, panning, and tilting. Finally, given the specification for the key frames, the in-betweens can be automatically generated.

New Words & Expressions

animation	n. 动画	imply	v. 暗指，暗示
visual	a. 视觉的	scene	n. 场景，情景
translation	n. 平移	rotation	n. 旋转
variation	n. 变化	transparency	n. 透明性
surface texture	表面纹理	automobile	n.〈主美〉汽车
engine	n. 引擎	orientation	n. 方向，方位，定位
focal length	焦距	lighting effect	n. 光照效果
illumination	n. 照明	rendering	n. 着色，绘画
storyboard	n. 剧本	layout	n. 布置，版面安排
key-frame	关键帧	in-between frames	n. 插值帧
frame-by-frame	a. 逐帧的	consecutively	adv. 连续地，连贯的

Notes

1. real-time: 实时的；produced by flight simulators: 过去分词短语做定语，修饰 animations。for instance: 插入语。全句译为：飞行模拟器生成的实时计算机动画按飞机控制器上的动作来显示动画序列。
2. in "real-time playback" mode: 以实时回放的模式。

英文摘要的写作技巧

英文摘要（Abstract）的写作应用很广。不仅参加国际学术会议、向国际学术刊物投稿要写摘要，国内级别较高的学术期刊也要求附上英文摘要。学位论文更是如此。论文摘要是全文的精华，是对一项科学研究工作的总结，对研究目的、方法和研究结果的概括。

一、摘要的种类与特点

摘要主要有以下四种。

第一种是随同论文一起在学术刊物上发表的摘要。这种摘要置于主体部分之前，目的是让读者首先了解一下论文的内容，以便决定是否阅读全文。一般来说，这种摘要在全文完成之后写。字数限制在100~150字之间。内容包括研究目的、研究方法、研究结果和主要结论。

第二种是学术会议论文摘要。会议论文摘要往往在会议召开之前几个月撰写，目的是交给会议论文评审委员会评阅，从而决定是否能够录用。所以，比第一种略为详细，长度在200~300字之间。会议论文摘要的开头有必要简单介绍一下研究课题的意义、目的、宗旨等。如果在写摘要时，研究工作尚未完成，全部研究结果还未得到，那么，应在方法、目的、宗旨、假设等方面多花笔墨。

第三种为学位论文摘要。学士、硕士和博士论文摘要一般都要求用中、英文两种语言写。学位论文摘要一般在400字左右，根据需要可以分为几个段落。内容一般包括研究背景、意义、主旨和目的；基本理论依据，基本假设；研究方法；研究结果；主要创新点；简短讨论。不同级别的学位论文摘要，要突出不同程度的创新之处，指出有何新的观点、见解或解决问题的新方法。

第四种是脱离原文而独立发表的摘要。这种摘要更应该具有独立性、自含性、完整性。读者无需阅读全文，便可以了解全文的主要内容。

二、摘要的内容与结构

摘要内容一般包括：
- 目的（objectives，purposes）：包括研究背景、范围、内容、要解决的问题及解决这一问题的重要性和意义。
- 方法（methods and materials）：包括材料、手段和过程。
- 结果与简短讨论（results and discussions）：包括数据与分析。
- 结论（conclusions）：主要结论，研究的价值和意义等。

概括地说，摘要必须回答"研究什么""怎么研究""得到了什么结果""结果说明了什么"等问题。

无论哪种摘要，语言特点和文体风格也都相同。首先必须符合格式规范。第二，语言必须规范通顺，准确得体，用词要确切、恰如其分，而且要避免非通用的符号、缩略语、生偏词。另外，摘要的语气要客观，不要做出言过其实的结论。

三、学术期刊论文摘要

清华大学学报（自然科学版）就如何写好科技论文英文摘要、提高 Ei 收录率，概括了论文摘要的写作规则，值得一读：

1. 摘要的目的

摘要是论文的梗概，提供论文的实质性内容的知识。摘要的目的在于：给读者关于文献内容的足够的信息，使读者决定是否要获得论文。

2. 摘要的要素

1）目的——研究、研制、调查等的前提、目的和任务，所涉及的主题范围。
2）方法——所用的原理、理论、条件、对象、材料、工艺、结构、手段、装备、程序等。
3）结果——实验的、研究的结果、数据，被确定的关系，观察结果，得到的效果、性能等。
4）结论——结果的分析、研究、比较、评价、应用，提出的问题等。

3. 摘要的篇幅

摘要的篇幅取决于论文的类型。但无论哪一种论文，都不能超过 150 words。可采用以下方法使摘要达到最小篇幅：

1）摘要中第一句的开头部分,不要与论文标题重复。
2）把背景信息删去，或减到最少。
3）只限于新的信息。过去的研究应删去或减到最小。
4）不应包含作者将来的计划。
5）不应包含不属于摘要的说法，如："本文所描述的工作，属于……首创""本文所描述的工作，目前尚未见报道""本文所描述的工作，是对于先前最新研究的一个改进"等。
6）相同的信息不要重复表达。如 at a temperature of 250℃ to 300℃，应改为 at 250~300℃；

at a high pressure of 1.2 MPa，应改为 at 1.2 MPa。

7）以量的国际单位符号表示物理量单位（例如，以"kg"代替"kilogram"）。

8）以标准简化方法表示英文通用词（以"NY"代替"New York"）。

9）删去不必要的短语，如："A method is described""In this work""It is reported that""This paper is concerned with""Extensive investigations show that"等。

4. 摘要的英文写作风格

要写好英文摘要，就要完全地遵从通行的公认的英文摘要写作规范。其要点如下：

1）句子完整、清晰、简洁。

2）用简单句。为避免单调，改变句子的长度和句子的结构。

3）用过去时态描述作者的工作，因它是过去所做的。但是，用现在时态描述所做的结论。

4）避免使用动词的名词形式。如：

正："Thickness of plastic sheet was measured"

误："measurement of thickness of plastic sheet was made"

5）正确地使用冠词，既应避免多加冠词，也应避免蹩脚地省略冠词。如：

正："Pressure is a function of the temperature"

误："The pressure is a function of the temperature"；

正："The refinery operates …"

误："Refinery operates…"

6）使用长的、连串的形容词、名词、或形容词加名词来修饰名词。为打破这种状态，可使用介词短语，或用连字符连接名词词组中的名词，形成修饰单元。例如：

应写为"The chlorine-containing propylene-based polymer of high melt index"，而不写为"The chlorine containing high melt index-propylene based polymer"。

7）使用短的、简单的、具体的、熟悉的词。不使用华丽的词藻。

8）使用主动语态而不使用被动语态。"A exceeds B"读起来要好于"B is exceeded by A"。使用主动语态还有助于避免过多地使用类似于"is""was""are"和"were"这样的弱动词。

9）构成句子时，动词应靠近主语。避免形如以下的句子：

"The decolorization in solutions of the pigment in dioxane, which were exposed to 10 hr of UV irradiation, was no longer irreversible."

改进的句子，应当是：

"When the pigment was dissolved in dioxane, decolorization was irreversible, after 10 hr of UV irradiation."

10）避免使用那些既不说明问题，又没有任何含意的短语。例如："specially designed or formulated""The author discusses""The author studied"应删去。

11）不使用俚语、非英语的句子，慎用行话和口语，不使用电报体。

5. 学术期刊论文摘要实例

Optimization of Electromagnetic Devices
Using Intelligent Simulated Annealing Algorithm

Abstract —— Intelligent simulated annealing algorithm for the optimal design of electromagnetic devices is presented in this paper.[目的] The algorithm is implemented by assembling fuzzy inference into improved simulated annealing algorithm, which makes SA algorithm has the ability to reject infeasible solutions prior to objective function computation.[方法，结论] The algorithm is applied to the optimal design of a brushless dc motor and about 60 percents CPU time of SA can be saved.[结果] —— IEEE Transactions on Magnetics, September 1998

四、学位论文摘要

学位论文摘要也叫内容提要（summary）。一般单独占一页，装订在学位论文目录之前。这种摘要也是介绍研究背景、内容、目的、方法、结果等。但是，学位论文摘要与科技论文摘要的不同之处是，它必须指出研究结果的独到之处或创新点。关于研究的内容也可以稍加详细介绍，摘要的长度一般在 400 字左右。如有必要，可以分为几个段落。学位论文摘要可以分以下四步写。

开头部分可以先介绍一下研究的背景、宗旨、意义，提出问题，说明解决某一问题的必要性或重要性等。

第二步介绍本研究的目的、范围。

第三步介绍论文的主要内容。

最后对研究的主要内容，特别是新创造、新突破、新见解或新方法加以概括总结。当然，不同级别的学位论文（学士、硕士、博士），对创新性的衡量标准是不同的。

下面给出了一篇硕士学位论文摘要（略有改动），供读者参考。

基于 XML 的 Web 查询技术研究

WWW 是目前使用最为广泛的 Internet 信息服务系统，它为用户提供了一个搜索和浏览信息的工具。但 WWW 是一个信息的海洋，数亿万计的 Web 文档散布在世界各地成千上万台 Web 服务器上，且每台服务器自主管理各自的资源，没有一个统一的管理机制。总体状况是，整个网络上资源丰富，内容庞杂，很难实现数据的共享，亦不能进行有效的查询。（介绍研究背景，提出问题）

XML（Extensible Markup Language 可扩展标记语言）的出现为上述问题提供了理想的解决方案。XML 是特别为 Web 应用设计的，是一个在互联网上进行数据交换的理想工具。

由于 XML 能使不同来源的结构化数据很容易地结合在一起，从而使搜索多样化的不均匀数据成为可能，为解决 Web 查询问题带来了希望。本文围绕着基于 XML 的 Web 查询技术涉及到的几个主要问题展开研究与讨论（第二、三段介绍研究的目的、范围）：

1. 针对 Web 上的数据格式差异大、数据来源差别大及当前的数据模型不能很好地描述 Web 上数据的缺点，研究和探讨了 Web 上的数据模型——XML 数据模型。

2. 解决异构数据的集成问题。由于异构数据的集成主要是异构关系数据库中数据的集成。所以本文采用了基于"中间模式"的方法进行数据的集成，将关系数据库中的数据转换为通用的 XML 格式的数据，从而更好地实现数据的共享。

3. 利用 XML 数据模型在半结构化数据表示和查询方面的优势,通过基于 XML 的查询语言 XML-QL 实施 Web 查询（第四、五、六段介绍论文的主要内容）。

在 XML 出现以前，面向 Web 的数据查询是一项复杂的技术。通过本文研究的方法，如果采用 XML 数据模型描述 Web 上的数据,把不同来源的结构化数据转换成 XML 格式的数据，就能够方便地实现数据的共享和进行精确、有效的信息查询，将极大地简化复杂性，提高工作

效率。(对研究的主要内容进行概括，说明研究的意义)

关键词：XML；Web 查询；半结构化数据；数据模型

WWW is the most widely used information service system on the Internet. It offers a tool to search and browse information. But WWW contains too much information and billions of pages located on thousands of servers in different palaces in the world. Every server manages its own resources, and does not follow one standard managing mechanism. In general, there are wealthy of various resources on the net, which cannot be shared and queried effectively.

The emergence of XML (Extensible Markup Language) provides a perfect solution to the above-mentioned problem. XML, specially devised for Web application, is an excellent tool for data exchange on the Internet.

With its great capability to combine structured data with different sources, XML makes it possible to retrieve data from different and incompatible database servers, and brings hopes for Web data query. This paper focus on the following problems:

1. Considering the striking difference of data format and data sources on the Web, we made research and discuss on the Web data model——XML data model.

2. Solve the problem of integrating data with different structure. Because the integration problem mainly focus on the integration of different structured RDBES, this article adopted mediated schema and transformed the data in RDBMS into XML data source. Therefore, better sharing of data can be realized.

3. Making full use of the advantages of XML data model in semi-structured data representation and query, we carry out data retrieving queries from XML data resources on Web by using XML-QL.

Before the emergence of XML, Web oriented data query is a complex technique. If we represent Web data by XML data model and transform the structured data from different data source into XML data format sources as discussed in this paper, we will realize data sharing and carry out exact and efficient data query, which will simplify our work and enhance efficiency greatly.

Key Words: XML, Web query, Semi-structured data, Data model

Exercises

I. Answer the following questions

1. How many video formats are there in which the computer can use to present image? And what are they?
2. How many audio formats are there in which the computer can use to present sound? And what are they?
3. Except education and commercial uses, what other multimedia applications can you imagine?
4. Generally, how to design an animation sequence?
5. What are the in-between frames?

Chapter 13　Anatomy of the Internet of Things

学习指导

　　物联网是新一代信息技术的重要组成部分，也是"信息化"时代的重要发展阶段。顾名思义，物联网就是物物相连的互联网。这有两层意思：其一，物联网的核心和基础仍然是互联网，是在互联网基础上的延伸和扩展的网络；其二，其用户端延伸和扩展到了任何物品与物品之间，进行信息交换和通信，也就是物物相息。物联网通过智能感知、识别技术与普适计算等通信感知技术，广泛应用于网络的融合中，也因此被称为继计算机、互联网之后世界信息产业发展的第三次浪潮。

　　通过本章的学习，读者应掌握以下内容：
- 物联网的定义和核心；
- 物联网的协议及其与互联网协议的区别；
- 物联网的实现与应用。

It may appear to be a daunting task to engineer a new networking architecture for the Internet of Things (IoT). Yet nothing less than a completely new approach is needed. The Internet of Things environment is so different, and the devices to be connected so varied, that there has never been a networking challenge quite like it since the origin of what is now called the Internet.

In developing this new architecture for the Internet of Things, key lessons have been drawn from the development of the traditional Internet and other transformational technologies to provide some basic guiding principles:

- It should specify as little as possible and leave much open for others to innovate.
- Systems must be designed to fail gracefully: seeking not to eliminate errors, but to accommodate them.
- Graduated degrees of networking functionality and complexity are applied only where and when needed.
- The architecture is created from simple concepts that build into complex systems using the analog provided by natural phenomena.
- Meaning may be extracted from data in real time.

The emerging architecture for the Internet of Things is intended to be more inclusive of a wider variety of market participants by reducing the amount of networking knowledge and resources needed at the edges of the network. This architecture must also be extremely tolerant of failures, errors, and intermittent connections at this level. (Counter intuitively, the best approach is to *simplify* protocols at the edge rather than to make them more complex.)

In turn, increasing sophistication of networking capabilities are applied at gateways into the traditional Internet, in which propagator nodes provide communications services for armies of relatively unsophisticated devices.

Finally, meaning can be extracted from the universe of data in integrator functions that provide the human interface to the Internet of Things. This level of oversight is applied only at the highest level of the network; simpler devices, like worker bees in a hive, need not be burdened with computational or networking resources.

To explore what's needed for this new architecture, it is first necessary to abandon the networking status quo.

13.1 Traditional Internet Protocols Aren't the Solution for Much of the IoT

When contemplating how the Internet of Things will work, it helps to forget the conventional wisdom regarding traditional networking schemes—especially wide area networking (WAN) and wireless networking. In traditional WAN and wireless networking, the bandwidth or spectrum is expensive and limited, and the amount of data to be transmitted is large and always growing. Although over-provisioning data paths in wiring the desktop (and a majority of the traditional Internet) is commonplace, this isn't usually practical in the WAN or wireless network—it's just too expensive. With carriers largely bearing the cost and passing it along to customers, wireless costs range as high as ten times the wired equivalents using IP.

Besides cost, there's the matter of potential data loss and (in the wireless world) collisions. Traditional networking protocols include lots of checks and double-checks on message integrity to minimize costly retransmissions. These constraints led to today's familiar protocol stacks, such as TCP/IP and 802.11.

13.1.1 Introducing the "Chirp"

In most of the Internet of Things, however, the situation is completely different. The costs of wireless and wide-area bandwidth are still high, to be sure. And because many of the connections at the edge of the network—the IoT frontier, so to speak—will be wireless and/or lossy, any Internet of Things architecture must address these factors. But the amounts of data from most devices will be almost *immeasurably low* and the delivery of any single message completely uncritical. As discussed previously, the IoT is lossy and intermittent, so the end devices will be designed to function perfectly well even if they miss sending or receiving data for a while—even for a long while. As discussed earlier, it is this self-sufficiency that eliminates the criticality of any single message.

After reviewing all existing options in considering the needs of the IoT architecture from the ground up, it is clearly necessary to define a new type of data frame or packet. This new type of packet offers only the amount of overhead and functionality needed for simple IoT devices at the edge of the network—and no more. These small data packets, which are called chirps, are the fundamental building block of the emerging architecture for the IoT. Chirps are different from traditional Internet protocol packets in many ways (see the "Why Not the IP for the IoT?" sidebar. Fundamental characteristics of chirps include the following:

- Chirps incorporate only minimal overhead payloads, "arrows" of transmission (see below), simple *non-unique* addresses, and modest checksums.
- Chirps are inherently individually noncritical by design.
- Therefore, chirps include no retransmission or acknowledgment protocols.

Any additional functions necessary for carrying chirp traffic over the traditional Internet, such as global addressing, routing, and so on, are handled autonomously by other network devices by means of adding information to received simple chirps. There are therefore no provisions made for these functions within a chirp packet.

Lightweight and Disposable

In contrast to traditional networking packet structures, IoT chirps are like pollen or bird songs: lightweight, broadly propagated, and with meaning only to the "interested" integrator functions or end devices. The IoT is receiver-centric, not sender-centric, as is IP. Because IoT chirps are so small and *no individual chirp is critical*, there is limited concern over retries and resulting broadcast storms, which are a danger in IP.

It's true that efficient IoT propagator nodes will prune and bundle broadcasts (see Figure 13-1), but seasonal or episodic broadcast storms from end devices are much less of a problem because the chirps are small (and thus cause less congestion) and individually uncritical. Excessive chirps may thus be discarded by propagator nodes as necessary.

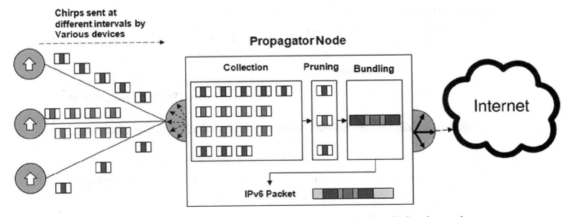

Fig.13-1 Chirps are typically collected within propagator nodes, bundled and pruned as necessary for transmission, and then typically forwarded via IPv6 over the traditional Internet

13.1.2 Functionality the IoT Needs—and Doesn't

This very different view of networking means that huge packets, security at the publisher, and assured delivery of any *single* message are unnecessary, allowing for massive networks based on extremely lightweight components. In one sense, this makes the IoT more "female" (receiver-oriented) than the "male" structure of IP (sender-oriented).

But there is obviously no point in having an IoT if nothing *ever* gets through. How can the acknowledged unpredictable nature of connections be managed? The answer, perhaps surprisingly, is over-provisioning—but only very locally between chirp device and propagator node. That is, these short, simple chirps may be re-sent over and over again as a brute-force means of ensuring that some get through.

13.1.3 Efficiency Out of Redundancy

As seen in Figure 13-2, because the chunks of data are so small, the costs of this over-provisioning at the very edge of the IoT are infinitesimal. (They are often handled by local Wi-Fi, Bluetooth, infrared, and so on, so they are not metered by any carrier.) Therefore, the benefits of this sort of scheme are huge. Because no individual message is critical, there's no need for any error-recovery or integrity-checking overhead (except for the most basic checksum to avoid a garbled message). Each chirp message simply has an address, a short data field, and a checksum. In some ways, these messages are what IP datagrams were meant to be. Chirps are also similar in many ways to the concepts of the Simple Network Management Protocol (SNMP), with simple "get" and "set" functionality.

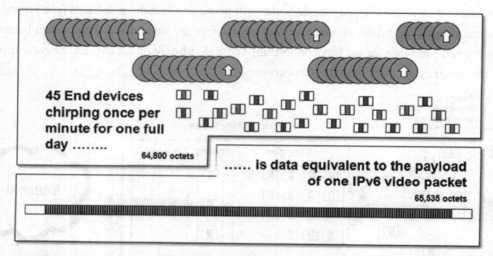

Fig.13-2   Many small chirps (machine-to-machine–oriented) are still considerably lessdata than a much longer IP packet (human-oriented)

Importantly, the cost and complexity burden on the end devices to incorporate chirp messaging will be very low–because it must be in the IoT. The most efficient integration schemes will likely be "chirp on a chip" approaches, with minimal data input/output and transmission/reception functionality combined in a simple standardized package.

The chirp will also incorporate the "arrow" of transmission mentioned previously, identifying the general direction of the message: whether toward end devices or toward integrator functions (see Figure 13-3). Messages moving to or from end devices need only the address of the end device; where it is headed or where it is from is *unimportant* to the vast majority of simple end devices.

These devices are merely broadcasting and/or listening, and local relevancy or irrelevancy is all that matters.

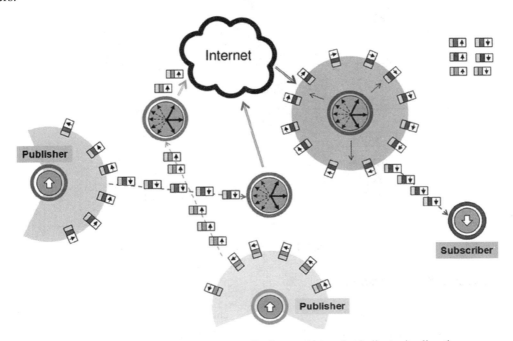

Fig.13-3　Each chirp includes an "arrow" of transmission that indicates its direction of propagation: toward end devices or toward integrator functions

So the end devices may be awash in the ebb and flow of countless transmissions. They may broadcast continuously and trust that propagator nodes and integrator functions elsewhere in the network will delete or ignore redundant messages. Likewise, they may receive countless identical messages before detecting one that has changed and requires an action in response.

In essence, this means that the chirp protocol is "wasteful" in terms of retransmissions only very locally, where bandwidth is cheap or free (essentially "off the net"). But because propagator nodes are designed to minimize the amount of superfluous or repeated traffic that is forwarded, WAN costs and traffic to the traditional Internet are vastly reduced.

Note that, unlike traditional network end devices such as smartphones and laptops, the largest percentage of IoT end devices likely *will not* include both send and receive functions (see Figure 13-4). An air quality sensor, for example, needs to send only the current state for whatever chemicals it is measuring. It begins sending when powered on, and repeatedly chirps this information until switched off. This may simplify significantly the hardware and embedded software needed at the vast majority of end points.

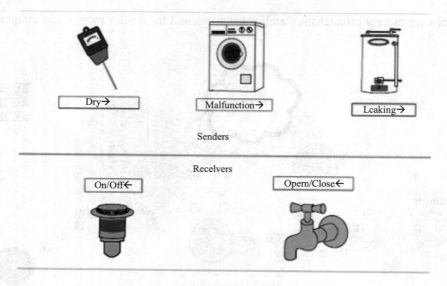

Fig.13-4　Many IoT devices will be send-only or receive-only

New Words & Expressions

accommodate　vt. 容纳；使适应
intermittent　a. 间歇的；断断续续的
propagate　vt. 传播，宣传
chunk　n. 相当大的量
ebb　n. 退潮，衰落
prune　vt. 精简某事物；除去某事物多余的部分
congestion　n. 拥挤，阻塞
dauntin　adj. 令人畏惧的；使人气馁的；令人怯步的

chirp　n. [通信] 啁啾声，一种脉冲编码技术
pollen　n. 花粉
episodic　a. 由松散片段组成的，插曲般的
infrared　n. 红外线
intuitively　adv. 直觉地；直观地
bundle　vt. & vi. 收集，归拢
discard　vt. 丢弃，抛弃

Abbreviations:

IoT(Internet of Things)　物联网
WAN (wide area networking) 广域网
SNMP (Simple Network Management Protocol)

13.2　It's All Relative

The detailed structure of the chirp packet is not discussed in this Chapter, but a brief introduction is useful here. The key difference between the Internet of Things packet and other packet formats is that the meanings of values within the packet are relative. That is, there is no fixed definition for the packet locating headers, addresses, and so on (as there is for IPv6, for example).

As seen in Figure 11-3, *markers* are used in place of a fixed format definition to allow receiving devices to determine information such as sending address, type of sensor and data, arrow of transmission, and so on. These markers are both public and private types.

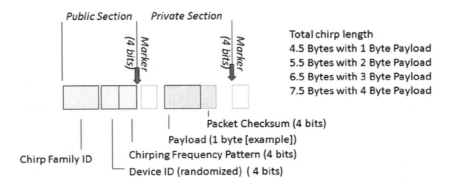

Fig.11-5 The IoT chirp packet is unique in that addressing and other information is determined by relative position to defined markers, not by a rigid general overall protocol formats

Public markers, which are found in every IoT packet, allow the receiving device to "parse" the incoming traffic. When a public marker is noted, the receiving device examines data ahead of and behind the marker for specific bits needed to determine how the rest of the packet will be forwarded and/or acted upon. The receiving device need not examine the packet except for the areas indicated by the location and type of public marker observed. Public markers include the basic arrow of transmission described previously, a limited 4-bit checksum for packet verification, and so on. Bits in the data field that are not part of the routing and verification information are simply treated as a data payload at this level of examination.

13.2.1 Format Flexibility

The presence of public markers within the IoT chirp packet permits the length of the IoT packet to vary as necessary for the specific application, device type, or message format. Different families of IoT packets with varying amounts of public data fields are defined to allow sufficient information to be added for applications that need additional context, but also to allow for minimal overhead for the most basic device types and generic IoT packet propagation.

The use of public markers is inspired by nature, including the transcription or "reading" of heredity information coded in DNA within genes to create proteins needed for development and life. DNA strands may contain repetitions and "junk" sections that should not be read, but localized markers are used to indicate "start" and "stop" points for transcription. Receiving devices use public markers in the same way to examine IoT chirp packets without requiring specific byte counts or other overhead-generating restrictions.

13.2.2 Private Markers for Customization and Extensibility

Private markers are permitted within the generic "data" field defined by public markers to allow customization of data formats for specific applications, manufacturers, and so on. As with public markers, the private marker allows a receiving device to parse the data stream to locate information for specific needs.

13.2.3 Addressing and "Rhythms"

As noted earlier, billions of end devices of the IoT will be extremely inexpensive and may be manufactured by makers throughout the world, many of whom will not have extensive networking knowledge. For this reason, ensuring address uniqueness through a centralized database of device addresses for the hundreds of billions of IoT end points is a nonstarter.

Part of the public information in the IoT chirp packet will be a simple, non-unique, 4-bit device ID applied through PC board traces, hardware straps, DIP switches, or similar means. As described in Chapter 8, it will combine with a randomly generated 4-bit pattern to ensure a much lower potential for two end devices, connected to the same local propagator node, to have identical identifications. (This combination of bits is also used to vary transmission rates in wireless environments to avoid a "deadly embrace.")

If additional addressing specificity and/or security is required in particular applications, it will be possible to add this information within the private space of the IoT packet.

13.2.4 Family Types

The final public information contained in all IoT chirp packets is a classification into one of 255 possible chirp "families." As described in Chapter 8, these families will primarily divide along type and application lines, such as sensors of various types, control valves, green/yellow/red status indicators, and so on. These chirp families will be defined from generic to more specific, and will be broad and extensible enough to allow any type of IoT application. As noted previously, for specific applications or devices in which more granularity of type classification is desired, this custom information may be defined by private markers within the data field.

The type and classification of the chirp packets enables one of the most far-reaching benefits of the IoT: the ability for data analyzers to discover and recruit new data sources based on affinities with information neighborhoods. Because this type and classification information is "external", it may be recognized and acted upon by many IoT elements, such as integrator functions and propagator nodes (along with their associated publishing agents, if so-equipped).

In this way, integrator functions monitoring a pressure sensor in a pipeline might seek out nearby temperature sensors to look for correlations that might provide richer information. The type and classification of the chirp packet alone conveys some potential knowledge that may be analyzed and coordinated with other information, and this is carried throughout the network as chirp packet streams are forwarded.

This feature is true even if the transmitting sensors were installed for a different application, by a different organization, or at a different point in time. The option for "public" advertising of type and classification allow broader use (and re-use) of chirp streams, by enabling dynamic publish/subscribe relationships to be created and modified over time as the IoT "learns."

This benefit is achieved without burdening end devices. Because most end devices are by definition very simple in the Internet of Things, those designed to receive IoT chirp packets will be required to process only the most basic of elements of the protocol (for example, using public markers to identify packets addressed to themselves and reading only that data). The IoT elements making much more extensive use of the capabilities of the chirp packet are those that must route or analyze data from many end devices, specifically the propagator nodes and integrator functions.

New Words & Expressions

parse　v. 从语法上分析　　　　　　payload　n. 负载
checksum　n. 校验和　　　　　　　context　n. 语境；上下文
repetitio　n. 重复，反复　　　　　　hardware strap　硬件带
valve　n. 阀　　　　　　　　　　　affinity　n. 密切关系

13.3　Applying Network Intelligence at Propagator Nodes

As noted previously, replicating even this highly efficient chirp protocol traffic indiscriminately throughout the IoT would clearly choke the network, so intelligence must be applied at levels above the individual end devices. This is the responsibility of *propagator nodes*, which are devices that create an overarching network topology to organize the sea of machine-to-machine interactions that make up the Internet of Things.

Propagator nodes are typically a combination of hardware and software distantly similar to WiFi access points. They handle "local" end devices, meaning that they interact with end devices essentially within the (usually) wireless transmission range of the propagator node. They can be specialized or used to receive chirps from a wide array of end devices. Eventually, there would be tens or perhaps hundreds of thousands of propagator nodes in a city like Las Vegas. Propagator nodes will use their knowledge of adjacencies to form a near-range picture of the network. They will locate in-range nearby propagator nodes, as well as end devices and integrator functions either attached directly to or reached via those propagator nodes. This information is used to create the network topology: eliminating loops and creating alternate paths for survivability.

The propagator nodes will intelligently package and prune the various chirp messages before broadcasting them to adjacent nodes. Examining the public markers, the simple checksum, and the "arrow" of transmission (toward end devices or toward integrator functions), damaged or redundant messages will be discarded. Groups of messages that are all to be propagated via an adjacent node may be bundled into one "meta" message–a small data "stream"–for efficient transmission. Arriving "meta" messages may be unpacked and repacked.

Some classes of propagator nodes will contain a software publishing agent. This publishing agent interacts with particular integrator functions to optimize data forwarding on behalf of the integrator. Propagator nodes with publishing agents may be "biased" to forward certain information in particular directions based on routing instructions passed down from the integrator functions interested in communicating with a particular functional, temporal, or geographic "neighborhood" of end devices. It is the integrator functions that will dictate the overall communications flow based on their needs to get data or set parameters in a neighborhood of IoT end devices.

In terms of discovery of new end devices, propagator nodes and integrator functions will be again similar to traditional networking architectures. When messages from or to new end devices appear, propagator nodes will forward them and add the addresses to their tables (see Figure 13-6). Appropriate age-out algorithms will allow for pruning the tables of adjacencies for devices that go offline or are mobile and are only passing through.

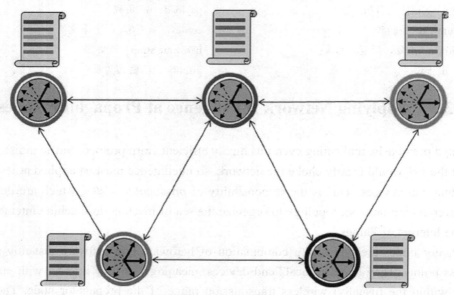

Fig.13-6　Propagator nodes independently build routing tables (and thus, the networktopology) based on the discovery of adjacent propagator nodes. Although not shown here,the location of integrator functions and discovered end devices would also be included inthe makeup of the topology

13.3.1　Transport and Functional Architectures

The emerging architecture of the Internet of Things combines two completely independent network topologies or architectures: *transport* and *functional*, as shown in Figure 13-7. The transport architecture is the infrastructure over which all traffic is moved and is provided primarily by propagator nodes (and the global Internet). The functional architecture is the virtual "zone" or "neighborhood" of interest created by integrator functions independent of physical paths.

The transport network portion of the Internet of Things operates with little or no context of the actual significance of the data chirps being handled. As noted previously, propagator nodes build the transport network based on more-traditional networking concepts and routing algorithms. End chirp devices may link to propagator nodes in a wide variety of ways: wirelessly via radio or optical wavelengths (see the following "Chirps in a Wireless World" sidebar), power line networking, a direct physical connection, and so on. A single propagator node can be connected to a large number of chirp devices and provide services for all. Unless the propagator node is biased by the integrator function, the basic model is "promiscuous forwarding."

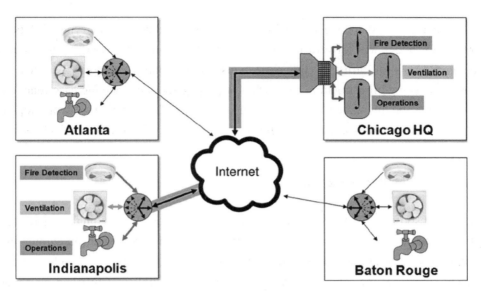

Fig.3-7　The network topology and logical topology of the Internet of Things can vary considerably

　　Propagator nodes bundle and convert chirp traffic as necessary for transport to adjacent propagator nodes and thence to integrator functions or chirp devices. The link between propagator nodes is typically a traditional networking protocol such as TCP/IP, but it can also be chirp-based.

　　Besides transporting the very simple chirps, the higher-level protocol packets created by the propagator nodes include additional contextual information not found in the chirps. This data may include additional address information related to location, time of day, and other factors, as shown in Figure 13-8. Thus, the propagator nodes increase the utility of the chirp data stream without burdening the vast numbers of end devices with networking cost and complexity. This additional contextual information is added only by propagator nodes and analyzed by integrator functions.

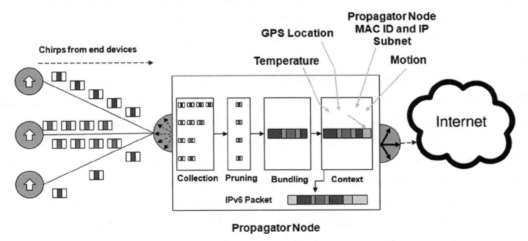

Fig.13-8　As chirps are bundled within propagator nodes, additional location, addressing, protocol, and other information is added

An important difference between the IoT transport architecture and many forms of traditional

Chapter 13　Anatomy of the Internet of Things

networking is that it is fundamentally egalitarian, similar to wind currents carrying all types of plant pollen. Propagator nodes will forward IoT traffic to and from *any* end device or integrator function within the constraints of "trust" "communications" and "control" factors. The IoT can then "piggyback" on existing infrastructure, and each new propagator node may increase functionality for a variety of users and integrator functions. Fundamentally, the transport network topology and architecture does not create (or limit) the *functional* IoT network topology, which is created by integrator functions.

13.3.2 Functional Network Topology

With the transport network architecture (described previously) providing forwarding services for chirps in both directions ("down" toward chirp devices and "up" toward integrator functions), attention may now be turned to the functional IoT architecture, which is overlaid on the transport architecture in somewhat the same way that the propagation of pollen is overlaid on general wind currents in the atmosphere.

The functional network of the IoT, then, becomes less a matter of how the "wires" (physical or virtual) are connected and much more a matter of information that is of interest. The emerging architecture of the Internet is fundamentally a "publish and subscribe" model driven by the integrator functions. It is also receiver-oriented, with the machine at the far end of the transmission "arrow" determining what data is pertinent and useful.

13.3.3 Defined by Integrator Functions

At this point, a brief description of the integrator function is appropriate. Integrator functions may take a wide array of physical forms, and multiple logical integrator functions can be deployed on one machine with a single connection to the traditional Internet (perhaps via a filter gateway). From a functional standpoint, they are somewhat autonomous creators of relationships with a select group of end points.

As an example, imagine an integrator function designed to monitor moisture content in the far-flung fields of an agribusiness concern (see Figure 13-9). The moisture–sensing end devices broadcast chirps at intervals, indicating the moisture content of the surrounding soil. The tiny chirps of data have a transport "arrow" pointing toward integrator functions.

The chirps are received by in-range propagator nodes deployed by the agribusiness concern (or anyone else). As noted previously, these chirps leave the propagator nodes bundled with additional contextual information such as a full IPv6 address and location information, allowing a more precise location and identification of the specific individual sensor that is not available from the simple chirp. The transport network of the propagator node essentially "publishes" these data streams via the traditional Internet.

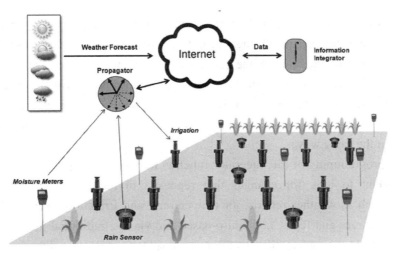

Fig.13-9 An integrator function retrieves data from end devices such as moisturesensors and external feeds such as the expected precipitation and humidity, using the information to control irrigation valves

13.3.4 Harvesting Information from the IoT

The preceding description suggests a virtual private sensor network, with a single agribusiness supplier installing its own end-device sensor propagator nodes, using the traditional Internet to create a routing path, and then monitoring the network privately for its own benefit. And certainly many IoT big data "neighborhoods" are created in this way. But there is also a tremendous potential for building networks that rely on data provided by Internet of Things elements not owned, managed, and controlled by a single source.

In the emerging social networking culture in the Western world, crowd sourcing and data sharing is becoming more commonplace. In light of this, individuals and organizations may choose to install sensors, cameras, and other devices of all kinds locally, providing the IoT streams from these devices generically and publicly. (Note that many individuals and groups do this today with web cams, weather sensors, and the like using traditional Internet protocols such as IP).

Propagator nodes set to promiscuously forward generic chirps would simply move these packets in the general direction of integrator functions. (Note that it is possible for propagator nodes to be used for both private and public streams simultaneously—offering transport for the general good, as it were.)

An integrator function might be configured, then, to gather data from interesting end devices that it has discovered by searching out small data streams from specific classes of device, location, or other characteristics. These integrator functions might combine small data streams from many independent end devices installed by any number of unknown individuals to create interesting new big data information.

13.3.5 Programming and "Bias"

Human programming of the integrator function may instruct it to look for certain locations and types of data streams via the traditional Internet, or the integrator functions may identify potentially interesting candidate data streams through affinities with known sources. Locating appropriate moisture sensor streams on the Internet, the integrator function begins to receive and incorporate this data. The integrator function may even "bias" the publishing agent within propagator nodes (if so-equipped) for some efficiency in combining chirps into larger packets in small data streams or discarding duplicate chirps. (Attached filter gateways might also serve to prune and select from verbose streams in the same way.

The human programming of the integrator function may now incorporate these streams of data on moisture content to look for changes that represent drying out beyond preset thresholds. Additional data, such as weather reports, air temperature, and irrigation reservoir levels (acquired from a variety of sources and feeds, both chirp-based and via the traditional Internet), might also be incorporated to provide a complete picture of irrigation needs for current and future periods of time.

The resulting reports might be provided for human action. Or, in a more automated scenario, the integrator function might respond (via its programming) to change watering times or durations in specific fields (if irrigation valves are also under IoT control). In this application, the integrator function might also analyze video surveillance streams to confirm that sprinklers are on and operating normally.

Note that this functional IoT network might interconnect over any transport topology. The agribusiness need not build out its own private network for the entire transport path; instead, it can use the traditional Internet for much of the transport infrastructure. The enterprise might deploy only the moisture sensors and some specialized propagator nodes, as appropriate.

This is only one example out of millions that might be imagined for the Internet of Things. But the basic principles of very simple devices at the edge, publish-and-subscribe, utilization of public network transport, and integration of a variety of data sources apply broadly.

13.3.6　Receiver-Oriented Selectivity

In the same way that female plants "select" only the appropriate pollen from the same species and reject foreign pollen, dust, and other material, integrator functions are similarly selective in choosing which chirp streams to incorporate as inputs for analysis. Integrator functions may be programmed to "set," configure, or otherwise manipulate end devices by generating "chirp" traffic of their own that is packaged for routing through the traditional Internet to a propagator known to be near the target end device. With the transmission "arrow" set in the direction of end devices, these packets are transported to the appropriate propagator node (typically within IPv6 packets) and then output as IoT chirps. Integrator functions may combine chirps for widely scattered end devices in a single broadcast packet, which is then pruned and rebroadcast as necessary by intermediate propagator nodes.

The end devices may be able to "hear" a variety of traffic, but thanks to similar receiver-oriented selectivity, they act upon only the specific traffic intended for them. As noted earlier, the intermediate routing and addressing information is primarily a function of the propagator nodes; end devices need only detect the simple IoT chirp addresses.

New Words & Expressions

replicat v. 复制，模拟
prune v. 修剪，削减
temporal adj. 时间的；暂存的
adjacency n. 邻接
meta pref. 介于…之间
piggyback vt. 背，附带发生

求职英语简介

一、个人简历

个人简历（resume / curriculum vitae）是升学、求职过程中的重要文件。英文个人简历的格式比较固定，一般应包括个人信息、联系地址、求职愿望、学历、工作经历、证明人等部分，也可以根据具体情况，适当增加有关条目。内容的安排要求清晰易读、层次分明，学历和工作经历一般可从最近开始，逐渐向后追述。从语言方面来看，个人简历多以名词词组为主，也可以使用动词词组，但是要求结构整齐对称。

以下是某位学生的个人简历，他申请的工作职位是某计算机公司的售后服务技术员。一般来说，他还需另外撰写一份求职信，介绍一下本人的基本情况、特长及对该公司业务的熟悉程度，同时重申个人的工作愿望，并提出面试请求。请大家阅读以下简历，注意其格式及语言特点，然后分析一下自己的情况，撰写一份个人简历。

Resume

Personal details
Name: Li Jian
Age: 20
Date of Birth: May 21,1981
Marital Status: single
Address: 36 Yanshan Street, Jinan, Shandong 250014, P. R. China
Tel: 0531-8937777
E-mail: lijian@hotmail.com

Position applied for After-sale Service Technician

Education
September 1995 to present   Shandong Institute of Electronic Technology,
Major in Applied Computing with special interest in networking & website implementation.
Courses taken: C Programming, Software Engineering, Visual Foxpro, Microsfot Office, Data Structures in C, English for Computing, Computer Networks, Internet Application, etc.

September 1992 to July 1995	Jinan No.14 Middle School, Jinan, Shandong. Monitor of class, outstanding in math & physics.

Work experience

Summer 1998	Company: ABC Computer Corp. Post: Sales Promotion Responsibility: organizing product exhibition tours & market survey.
Summer 1997	Company: Hope Co., Ltd Post: System Integration Responsibility: assembling PCs according to customer requirements; fixing hardware problems; on-the-spot installation & testing.
Summer, 1996	Company: Eastsoft Co., Ltd Post: Reception Responsibility: understanding customer needs; explaining Eastsoft products & services; highlighting Eastsoft after-sale service.
Honors	Top Prize in Homepage Design Contest, 1998.
Other information	Fluent in English with a TOEFL score of 590.
References	Mr.Liu Xinhua Ms. Wang Ying Sales Manager Manager of Customer Relations ABC Computer Corp. Eastsoft Co., Ltd 38 Heping Street 135 Wenhua Avenue Jinan, Shandong, 250014 Jinan, Shandong 250014 P. R. China P. R. China

二、求职申请信

求职信实质上是一种推销自己的商业函件，它通常与简历等一块儿寄出。其主要目的是说服雇主或招聘单位雇用你，或至少为你赢得面试的机会，是求职成功与否很关键的第一步。

求职可分为毕业后求职、跳槽求职和业余兼职三大类。无论是哪一种求职申请信，都应该认真撰写，做到简明扼要，重点突出，千万不要在信中罗列过多的琐碎事实，或对自己百般吹嘘炫耀，因为你在随信寄出的简历中已说明了最重要的资料。

一般来讲，求职信主要应包括以下几项内容：

（1）点明你所要申请的工作职位；

（2）说明你从何得知这份工作的信息（如从同事朋友处或从报纸的招聘广告等）；

（3）引起招聘者对简历中相关内容的注意；

（4）说明你是此项工作的最佳人选。

前两项通常在信函的第一段中开门见山加以说明。后两项是申请信的主要内容，放在第二段。本段表达成功与否起着很重要的作用，对求职者的能力和个性也是一种检验。在写作时，应指出你的简历中与工作特别相关的内容，以引起招聘者的注意和兴趣。在求职信的收尾段中，应说明联络方式并表达希望得到面试的机会。

求职申请信写好之后，应选用优质白纸打印出来。布局排版要美观大方，给人以整洁清新之感。求职申请信的格式与普通书信的格式基本一致。在寄发求职申请信时，应按招聘要求随函附上个人简历或和自传等有关材料。一个考虑周到的求职者还会随函附上一个已写好自己的通讯地址并贴好邮票的信封或明信片，供招聘单位复函之用。

总之，注意求职申请信从内容至形式的每一个细节，将为你赢得面试机会，最终找到称心如意的工作。下面是一篇毕业求职范例，供读者参考。

谋求证券经纪人之职 (Application for a Position of Stock Broker)

Gentlemen：

Your advertisement for stockbrokers in "China Daily" of May 2 has interested me very much. I feel I can fit one of the vacancies you have.

I am twenty-two years of age and female. I will graduate in July. My specialty at college is none other than securities business. I have gained some practical experience in brokerage in recent years acting as a part-time broker for Beijing Commodities Futures Exchange. At college, I have participated in a lot of activities both at the collegiate level and the departmental level. I have attended a large number of symposia and conferences on the business of stockbroking. I am young and energetic and maintain good interpersonal relations. I mention these because I know my line off work is a demanding one and I am willing to show that I can face up to any challenge placed before me.

Should this application meet with your favorable consideration, I will do my utmost to satisfy the confidence you may repose in me. I am looking forward to hearing from you as soon as possible.

Faithfully yours, Jing Jimei

Enclose: (1) My resume

(2) My academic record

(3) Two recent photos

本人对贵公司五月二日在《中国日报》上招聘证券经纪人的广告极感兴趣。我认为本人可以填补贵公司其中一个空缺的职位。

我现年22岁，女性，即将于7月份毕业。我在大学所学专业为证券交易。近两年来，我在北京商品期货交易所当兼职经纪人，从中获得了一些经纪人方面的实际经验。我在学校参加了许多活动，既有学校级的，也有高一级的。我还参加了不少以股票买卖为专题的研讨会和会议。本人精力充沛，而且有良好的人际关系。我之所以提及这些是因为我知道我想要从事的工作类型要求很严，但是我愿意面对出现在我面前的任何挑战。

贵公司如对我的申请惠予考虑，本人将竭诚工作，以不负贵公司之愿望。期望你们尽早回复。

附件：　（1）个人简历
　　　　（2）成绩单
　　　　（3）两张近照

Exercises

I. Answer the following questions

1. Why traditional Internet protocols aren't the solution for much of the IoT?
2. What are the key technologies to the IoT?
3. Please describe the features of Iot.

II. Fill in the blanks in each of the following

1. These small data packets, which are called _____, are the fundamental building block of the emerging architecture for the IoT.
2. In contrast to traditional networking packet structures, IoT _____ are like pollen or bird songs: lightweight, broadly propagated, and with meaning only to the "interested" integrator functions or end devices.
3. Unlike traditional network end devices such as smartphones and laptops, the largest percentage of IoT end devices likely *will not* include both _____ and _____ functions.
4. The key difference between the Internet of Things packet and other packet formats is that the meanings of values within the packet are _____.
5. The emerging architecture of the Internet of Things combines two completely independent network topologies or architectures: _____ and _____.

Chapter 14　Cloud Computing

学习指导

云计算为计算资源的供给提供了新的模式,这种模式把计算资源的位置转移到了网络,以减少与硬件和软件资源相关的管理成本。它代表了长久以来将计算作为一种实用工具的愿景,也就是规模化经济原理有助于有效地降低计算资源的成本。

要充分利用云托管数据存储系统,很好地了解云计算技术的各个方面是非常重要的。通过本章的学习,读者应掌握以下内容:
- 云计算的关键定义;
- 云计算的相关技术;
- 云计算的服务模式部署模型;
- 了解当前最先进的公共云计算平台。

Cloud computing technology represents a new paradigm for the provisioning of computing resources. This paradigm shifts the location of resources to the network to reduce the costs associated with the management of hardware and software resources. It represents the long-held dream of envisioning computing as a utility where the economy of scale principles help to effectively drive down the cost of computing resources. Cloud computing simplifies the time-consuming processes of hardware provisioning, hardware purchasing and software deployment. Therefore, it promises a number of advantages for the deployment of data-intensive applications, such as elasticity of resources, pay-per-use cost model, low time to market, and the perception of unlimited resources and infinite scalability. Hence, it becomes possible, at least theoretically, to achieve unlimited throughput by continuously adding computing resources if the workload increases.

To take advantage of cloud-hosted data storage systems, it is important to well understand the different aspects of the cloud computing technology. This chapter provides an overview of cloud computing technology from the perspectives of key definitions (Sect. 14.1), related technologies (Sect. 14.2), service models (Sect. 14.3) and deployment models (Sect. 14.4), followed by Sect. 14.5 which analyzes state-of-the-art of current public cloud computing platforms, with focus on their provisioning capabilities. Section 14.6 summarizes the business benefits for building software applications using cloud computing technologies.

14.1　Definitions

Cloud computing is an emerging trend that leads to the next step of computing evolution, building on decades of research in virtualization, autonomic computing, grid computing, and utility computing, as well as more recent technologies in networking, web, and software services. Although cloud computing is widely accepted nowadays, the definition of cloud computing has been arguable, due to the diversity of technologies composing the overall view of cloud computing. From the research perspective, many researchers have proposed their definitions of cloud computing by extending the scope of their own research domains. From the view of service-oriented architecture, Dubrovnik implied cloud computing as *"a service-oriented architecture, reduced information technology overhead for the end-user, greater flexibility, reduced total cost of ownership, on-demand services, and many other things"*. Buyya et al. derived the definition from clusters and grids, acclaiming for the importance of service-level agreements (SLAs) between the service provider and customers, describing that cloud computing is *"a type of parallel and distributed system consisting of a collection of interconnected and virtualized computers that are dynamically provisioned and presented as one or more unified computing resource(s) based on SLAs"*. Armbrust et al. from Berkeley highlighted three aspects of cloud computing including illusion of infinite computing resources available on demand, no up-front commitment, and pay-per-use utility model, arguing that cloud computing *"consists of the service applications delivered over the Internet along with the data center hardware and systems software that provide those services"*. Moreover, from the industry perspective, more definitions and excerpts by industry experts can be categorized from the perspectives of scalability, elasticity, business models, and others.

It is hard to reach a singular agreement upon the definition of cloud computing, because of not only a fair amount of skepticism and confusion caused by various technologies, but also the prevalence of marketing hype. For that reason, National Institute of Standards and Technology has been working on proposing a guideline of cloud computing. The definition of cloud computing in the guideline has received fairly wide acceptance. It is described as:

"a model for enabling convenient, on-demand network access to a shared pool of configurable computing resources (e.g., networks, servers, storage, applications, and services) that can be rapidly provisioned and released with minimal management effort or service provider interaction"

According to this definition, cloud computing has the following essential characteristics:

1. *On-demand self-service.* A consumer can unilaterally provision computing capabilities, such as server time and network storage, as needed automatically without requiring human interaction with each service's provider.
2. *Broad network access.* Capabilities are available over the network and accessed through standard mechanisms that promote use by heterogeneous thin or thick client platforms (e.g., mobile phones, laptops, and PDAs).
3. *Resource pooling.* The provider's computing resources are pooled to serve multiple consumers using a multi-tenant model, with different physical and virtual resources dynamically assigned and reassigned according to consumer demand. There is a sense of location independence in that the customer generally has no control or knowledge over the exact location of the provided resources but may be able to specify location at a higher

level of abstraction (e.g., country,state, or datacenter). Examples of resources include storage, processing, memory,network bandwidth, virtual networks and virtual machines.

4. *Rapid elasticity*. Capabilities can be rapidly and elastically provisioned, in somecases automatically, to quickly scale out and rapidly released to quickly scale in.To the consumer, the capabilities available for provisioning often appear to be unlimited and can be purchased in any quantity at any time.

5. *Measured Service*. Cloud systems automatically control and optimize resourceuse by leveraging a metering capability at some level of abstraction appropriateto the type of service (e.g., storage, processing, bandwidth, and active useraccounts). Resource usage can be monitored, controlled, and reported providing transparency for both the provider and consumer of the utilized service.

New Words & Expressions

virtualization   n. 虚拟化
prevalence   n. 流行，盛行
resource pooling   动态资源池
skepticism   n. 怀疑论，怀疑态度
on-demand self-service   按需自助服务
leverage   v. 利用

14.2 Related Technologies for Cloud Computing

Cloud computing has evolved out of decades of research in different related technologies from which it has inherited some features and functionalities suchas virtualized environments, autonomic computing, grid computing, and utility computing. Figure 14.1 illustrates the evolution towards cloud computing in hosting software applications. In fact, cloud computing is often compared to the following technologies, each of which shares certain aspects with cloud computing.

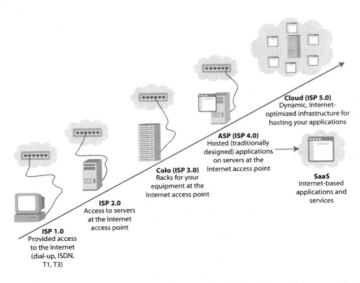

Fig. 14.1   The evolution towards cloud computing in hosting software applications

Table 14.1 provides a summary of the feature differences between those technologies and cloud computing in short, while details of related technologies are discussed as following:

Table 14.1 Feature similarities and differences between related technologies and cloud computing

Technologies	Differences	Similarities
Virtualization	Cloud computing is not only about virtualizing resources, but also about intelligently allocating resources for managing competing resource demands of the customers	Both isolate and abstract the low-level resources for high-level applications
Autonomic computing	The objective of cloud computing is focused on lowering the resource cost rather than to reduce system complexity as it is in autonomic computing	Both interconnect and intedgrate distributed computing systems
Grid computing	Cloud computing however also leverages virtualization to achieve on-demand resource sharing and dynamic resource provisioning	Both employ distribouted resources to achieve application-level objectives
Utility computing	Cloud computing is a realization of utility computing	Both offer better economic benefits

Virtualization

Virtualization is a technology that isolates and abstracts the low-level resources and provides virtualized resources for high-level applications. In the context of hardware virtualization, the details of physical hardware can be abstracted away with support of hypervisors, such as Linux Kernel-based Virtual Machine and Xen. A virtualized server managed by the hypervisor is commonly called a virtual machine. In general, several virtual machines can be abstracted from a single physical machine. With clusters of physical machines, hypervisors are capable of abstracting and pooling resources, as well as dynamically assigningor reassigning resources to virtual machines on-demand. Therefore, virtualization forms the foundation of cloud computing. Since a virtual machine is isolated from both the underlying hardware and other virtual machines. Providers can customize the platform to suit the needs of the customers by either exposing applications running within virtual machines as services, or providing direct access to virtual machines thereby allowing customers to build services with their own applications.Moreover, cloud computing is not only about virtualizing resources, but also about intelligent allocation of resources for managing competing resource demands of the customers. Figure 14.2 illustrates a sample exploitation of virtualization technology in the cloud computing environments.

Autonomic computing aims at building computing systems capable of self-management,which means being able to operate under defined general policies andrules without human intervention. The goal of autonomic computing is to overcome the rapidly growing complexity of computer system management, while being able to keep increasing inter connectivity and integration unabated. Although cloud computing exhibits certain similarities to automatic computing the way that it interconnects and integrates distributed data centers across continents, its objective some how is to lower the resource cost rather than to reduce system complexity.

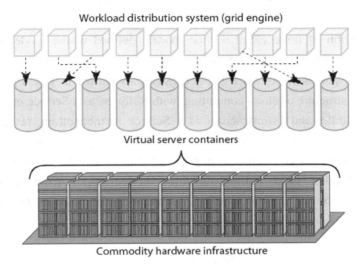

Fig. 14.2 Exploitation of virtualization technology in the architecture of cloud computing

Grid Computing

Grid computing is a distributed computing paradigm that coordinates networked resources to achieve a common computational objective. The development of grid computing was originally driven by scientific applications which are usually computation-intensive, but applications requiring the transfer and manipulation of a massive quantity of data was also able to take advantage of the grids.Cloud computing appears to be similar to grid computing in the way that it also employs distributed resources to achieve application-level objectives. However,cloud computing takes one step further by leveraging virtualization technologies to achieve on-demand resource sharing and dynamic resource provisioning.

Utility Computing

Utility computing represents the business model of packaging resources as ametered services similar to those provided by traditional public utility companies.In particular, it allows provisioning resources on demand and charging customers based on usage rather than a flat rate. The main benefit of utility computing is better economics. Cloud computing can be perceived as a realization of utility computing.With on-demand resource provisioning and utility-based pricing, customers are able to receive more resources to handle unanticipated peaks and only pay for resources they needed; meanwhile, service providers can maximize resource utilization and minimize their operating costs.

New Words & Expressions

evolve out of　从…发展而来
interconnectivity　n. 互联性
hypervisor　n. 管理程序（系统）
unabated　a. 不衰弱的，不减弱的

14.3　Cloud Service Models

The categorization of three cloud service models defined in the guideline are also widely accepted nowadays. The three service models are namely Infrastructure as a Service (IaaS), Platform as a Service (PaaS), and Software as a Service (SaaS).As shown in Fig. 14.3, the three service models form a stack structure of cloud computing, with Software as a Service on the top, Platform as a Service in the middle, and Infrastructure as a Service at the bottom, respectively. While the inverted triangle shows the possible proportion of providers of each model, it is worth mentioning that definitions of three service models from the guideline paid more attentions to the customers' view. In contrast, Vaquero et al. defined the three service models from the perspective of the providers'view. The following definitions of the three models combines the two perspectives, in the hope of showing the whole picture.

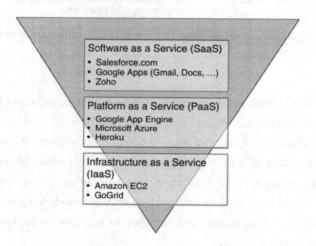

Fig. 14.3　The service models of cloud computing

1. *Infrastructure as a Service*: Through virtualization, the provider is capable of splitting, assigning, and dynamically resizing the cloud resources including processing, storage, networks, and other fundamental computing resources to build virtualized systems as requested by customers. Therefore, the customeris able to deploy and run arbitrary operating systems and applications. The customer does not need to deploy the underlying cloud infrastructure but hascontrol over which operating systems, storage options, and deployed applications to deploy with possibly limited control of select networking components. Thetypical providers are Amazon Elastic Compute Cloud (EC2) and GoGrid.
2. *Platform as a Service*: The provider offers an additional abstraction level, whichis a software platform on which the system runs. The change of the cloud resources including network, servers, operating systems, or storage is made ina transparent manner. The customer does not need to deploy the cloud resources,but has control over the deployed

applications and possibly application hosting environment configurations. Three platforms are well-known in this domain, namely Google App Engine, Microsoft Windows Azure Platform, and Heroku which is a platform built on top of Amazon EC2. The first one offers Python, Java, and Go as programming platforms. The second one supports languages in .NET Framework, Java, PHP, Python, and Node.js. While the third one is compatible with Ruby, Node.js, Clojure, Java, Python, and Scala.

3. Software as a Service: The provider provides services of potential interest to a wide variety of customers hosted in its cloud infrastructure. The services are accessible from various client devices through a thin client interface such as a web browser. The customer does not need to manage the cloud resources or even individual application capabilities. The customer could, possibly, be granted limited user-specific application configuration settings. A variety of services, operating as Software as a Service, are available in the Internet, including Salesforce.com, Google Apps, and Zoho.

New Words & Expressions

infrastructure   n. 基础设施；基础建筑
deploy   v. 部署，配置
arbitrary   a. 任意的
transparent   a. 透明的

14.4 Cloud Deployment Models

The guideline also defines four types of cloud deployment models, which are described as follows:

1. *Private cloud*. A cloud that is used exclusively by one organization. It may be managed by the organization or a third party and may exist on premise or off premise. A private cloud offers the highest degree of control over performance, reliability and security. However, they are often criticized for being similar to traditional proprietary server farms and do not provide benefits such as no up-front capital costs.
2. *Community cloud*. The cloud infrastructure is shared by several organizations and supports a specific community that has shared concerns (e.g., mission, security requirements, policy, and compliance considerations).
3. *Public cloud*. The cloud infrastructure is made available to the general public or a large industry group and is owned by an organization selling cloud services (e.g. Amazon, Google, Microsoft). Since customer requirements of cloud services are varying, service providers have to ensure that they can be flexible in their service delivery. Therefore, the quality of the provided services is specified using Service Level Agreement (SLA) which represents a contract between a provider and a consumer that specifies consumer requirements and the provider's commitment to them. Typically an SLA includes items such as uptime, privacy, security and backup procedures. In practice, Public clouds offer several key benefits to service consumers such as: including no initial capital investment

on infrastructure and shifting of risks to infrastructure providers. However, public clouds lack fine-grained control over data, network and security settings, which may hamper their effectiveness in many business scenarios.

4. *Hybrid cloud*. The cloud infrastructure is a composition of two or more clouds (private, community, or public) that remain unique entities but are bound together by standardized or proprietary technology that enables data and application portability (e.g., cloud bursting for load-balancing between clouds). In particular, cloud bursting is a technique used by hybrid clouds to provide additional resources to private clouds on an as-needed basis. If the private cloud has the processing power to handle its workloads, the hybrid cloud is not used. When workloads exceed the private cloud's capacity, the hybrid cloud automatically allocates additional resources to the private cloud. Therefore, Hybrid clouds offer more flexibility than both public and private clouds. Specifically, they provide tighter control and security over application data compared to public clouds, while still facilitating on-demand service expansion and contraction. On the down side, designing a hybrid cloud requires carefully determining the best split between public and private cloud components.

Table 14.2 summarizes the four cloud deployment models in terms of ownership, customership, location, and security.

Table 14.2 Summary of cloud deployment models

Deployment model	Ownership	Customership	Infrastructure locateon to customers	Secunity	Examples
Public cloud	Organization(s)	General public customers	Off-premises	No fine-grained control	Amazon Web Services
Private cloud	An organization/A third party	Customers within an organization	On/Off-premises	Highest degrec of control	Internal cloud platform to support business units in a large organization
Community cloud	Organization(s) in a community/A third party	Customers from organizations that have shared concerns	On/Off-premises	Shared control among organizations in a community	Healthcare cloud for exchanging health information among organizations
Hybrid cloud	Composition of two or more from above	Composition of two or more from above	On/Off-premises	Tighter control, but require careful split between distinct models	Cloud burstig for load balancing between cloud platforms

New Words & Expressions

premise n. 前提
uptime n. （计算机等的）正常运行时间
up-front a. 在前面的，预付的
portability n. 可移植性

14.5 Public Cloud Platforms: State-of-the-Art

Key players in public cloud computing domain including Amazon Web Services, Microsoft Windows Azure, Google App Engine, Eucalyptus, and GoGrid offer a variety of prepackaged services for monitoring, managing, and provisioning resources. However, the techniques implemented in each of these clouds do vary. For Amazon EC2, the three Amazon services, namely Amazon Elastic Load Balancer, Amazon Auto Scaling, and Amazon CloudWatch, together expose functionalities which are required for undertaking provisioning of application services on EC2. The Elastic Load Balancer service automatically provisionsin coming application workload across available EC2 instances while the AutoScaling service can be used to dynamically scale-in or scale-out the number of EC2 instances for handling changes in service demand patterns. Finally the CloudWatch service can be integrated with the above services for strategic decision making basedon collected real-time information.

Eucalyptus is an open source cloud computing platform. It is composed of three controllers. Among the controllers, the *cluster controller* is a key component that supports application service provisioning and load balancing. Each cluster controller is hosted on the *head node* of a cluster to interconnect the outer public networks and inner private networks together. By monitoring the state information of instances in the pool of server controllers, the cluster controller can select any available service/server for provisioning incoming requests. However, as compared to Amazon services, Eucalyptus still lacks some of the critical functionalities, suchas auto scaling for its built-in provisioner.

Fundamentally, Microsoft Windows Azure *fabric* has a weave-like structure,which is composed of node including servers and load balancers, and edges including power and Ethernet. The *fabric controller* manages a *service node* througha built-in service, named Azure Fabric Controller Agent, running in the background,tracking the state of the server, and reporting these metrics to the controller. If a fault state is reported, the controller can manage a reboot of the server or amigration of services from the current server to other healthy servers. Moreover,the controller also supports service provisioning by matching the VMs that meetrequired demands.

GoGrid Cloud Hosting offers developers the F5 Load Balancer for distributin gapplication service traffic across servers, as long as IPs and specific ports of these servers are attached. The load balancer provides the round robin algorithmand least connect algorithm for routing application service requests. Additionally,the load balancer is able to detect the occurrence of a server crash, redirecting furtherrequests to other available servers. But currently, GoGrid only gives developers aprogrammatic set of APIs to implement their custom auto-scaling service.

Unlike other cloud platforms, Google App Engine offers developers a scalable platform in which applications can run, rather than providing direct access to acustomized virtual machine. Therefore, access to the underlying operating systemis restricted in App Engine where load-balancing strategies, service provisioning,and auto scaling are all automatically managed by the system behind the scenes where the implementation is largely unknown. Chohan et al. have presented initial efforts of building App Engine-like framework, *AppScale*, on top of Amazon EC2 and Eucalyptus. Their offering consists of multiple components that automate deployment, management, scaling, and fault tolerance of an App Engine application. In their design and implementation, a single *AppLoadBalancer* existsin AppScale for distributing initial requests of users to the *AppServer*s of AppEngine applications. The users initially contact AppLoaderBalancer to request alogin to an App Engine application. The AppLoadBalander then authenticates the login and redirects request to a randomly selected AppServer. Once the request is redirected, the user can start contact the AppServer directly without going throughthe AppLoaderBalancer during the current session. The *AppController* sit inside the AppLoadBalancer is also in charge of monitoring the AppServers for growing and shrinking as the AppScale deployments happen over the time.

There is no single cloud infrastructure provider has their data centers at all possible locations through out the world. As a result, all cloud application providers currently have difficulty in meeting SLA expectations for all their customers. Hence,it is logical that each would build bespoke SLA management tools to provide better support for their specific needs. This kind of requirements often arises in enter prises with global operations and applications such as Internet service, media hosting, and Web 2.0 applications. This necessitates building technologies and algorithms for seamless integration of cloud infrastructure service providers for provisioning of services across different cloud providers.

New Words & Expressions

cluster controller　群集控制器　　　　　　　　bespoke　a. 预定的，定制的
seamless　a. 持续的；无缝的　　　　　　　　media hosting　媒体托管

14.6　Business Benefits of Cloud Computing

With cloud computing, organizations can consume shared computing and storage resources rather than building, operating, and improving infrastructure on their own. The speed of change in markets creates significant pressure on the enterprise IT infrastructure to adapt and deliver. In principle, cloud computing enables organizations to obtain a flexible and cost-effective IT infrastructure in much the same way that national electric grids enable homes and organizations to plug into a centrally managed, efficient, and cost-effective energy source. When freed from creating their own electricity, organizations were able to focus on the core competencies of their business and the needs of their customers. In particular, cloud computing technologies have provided some clear business benefits for building software applications. Examples of these benefits are:

1. *No upfront infrastructure investment*: Building a large-scale system may cost a fortune to invest in real estate, hardware (racks, machines, routers, backup power supplies), hardware management (power management, cooling), and operations personnel. Because of the high upfront costs, it usually takes several rounds of management approvals before the project could even get started. With cloud computing, there is no fixed cost or startup cost to start your project.

2. *Just-in-time Infrastructure*: In the past, if your system got famous and your infrastructure could not scale well at the right time, your application may becamea victim of its success. On the other hand, if you invested heavily and did not get famous, your application became a victim of your failure. By deploying applications in cloud environments, your application can smoothly scale as you grow.

3. *More efficient resource utilization*: System administrators usually worry about hardware

procuring (when they run out of capacity) and better infrastructure utilization (when they have excess and idle capacity). With cloud technology, they can manage resources more effectively and efficiently by having the applications request resources only what they need on-demand according to the *pay-as-you-go* philosophy.

4. *Potential for shrinking the processing time*: Parallelization is the one of the well-known techniques to speed up processing. For example, if you have a compute-intensiveor data-intensive job that can be run in parallel takes 500 hour to processon one machine. Using cloud technology, it would be possible to spawn andlaunch 500 instances and process the same job in 1 hour. Having available an elastic infrastructure provides the application with the ability to exploit parallelizationin a cost-effective manner reducing the total processing time.

New Words & Expressions

cost-effective   adj. 有成本效益的
excess   adj. 过剩的；过量的
spawn   vt. 大量生产

procure   vt. 取得，获得
parallelization   n. 并行化

广告文体简介

一、产品广告的结构 (Elements of product advertisement)

一则广告一般包含如下内容：
（1）标题，用于阐述该产品名称和最显著、最重要的特点，有的还用小标题进一步说明。
（2）产品样本照片或外形图、剖面图。这部分往往与标题相互配合，简短广告常省略此部分。
（3）正文列举其主要品种、规格、性能、用途和特点。
（4）商标、名称、代理机构、经销单位及地点、邮政编码、电话号码、用户电报、电报挂号及联系人等商业信息。

二、广告语言的特点 (Characteristics of advertising language)

广告与其他科技文体有很大区别，它的语言不但精练，而且生动、活泼。它广泛采用了形象、拟人、典故、夸张等修辞手法和押韵、叠文、一语双关等文字技巧，以增强广告的渲染力。广告英语具有如下特点。
（1）使用不同字体和图形，突出最重要的、最令人感兴趣的信息。这种手段简洁易懂，使人一目了然。如：
(a) Finally, the dBASE, you've been waiting IV.
　　Get the new dBASE IV, Now for just $ 449
这是一份计算机软件广告。它通过变换字体清晰地告诉用户"大家盼望已久的 dBASE 增强型版本 dBASE IV 终于问世了，该软件目前的售价仅为 449 美元。

(b) Twice the performance at half the cost.

The New Ultra Graphics Accelerator from Metheus.

"花一半的钱，却得双倍的性能，何乐而不为！"这是 Methus 公司为其图形加速器所作的广告。

（2）别出心裁，用违反常规的说法，夸大其辞。如：

Are you getting fat & lazy waiting for your plotter?

Don't wait! — Get PLUMP

"绘图仪的速度实在太慢，使人等得又胖又懒，还是买 PLUMP 吧！"这是一家公司为推销其 PLUMP 产品而作的极为夸张的广告。

（3）为使广告通俗易懂，常采用家喻户晓的口语和俗语等非正式文体，使公众感到亲切，便于记忆。如：

Here's proof that something small can be powerful.

这条广告说明了某微型机"体积虽小，但功能强大"。

A whole year without a single bug!

这说明该公司的产品质量过硬，"整整一年没有出一次故障"。

（4）为吸引顾客，使语言生动活泼，常采用形象化、拟人化等手段。例如，有一幅画着驴的广告，上写：

I feel like a donkey! For not buying the "Access" portable computer.

这是 Access 为其便携式计算机做的广告，它给人一种幽默感、新鲜感。

下面是两则广告，读者可从中体会广告文体的特点。

(a) Pentium, P6, P7: 86 Architecture's Attempt to Survive Future

As industry pundits put it, "No computer architecture can survive longer than two decades." Intel Corp, however, introduced Pentium in an attempt to extend the life of the 86 architecture which appears to be in its prime. Pentium achieves a performance comparable to RISC chips. The 86 architecture's survival strategy doesn't stop there. The next 86 family CPU, P6, will have more general-purpose registers, one of the 86 family's sore spots. The P6 will also offer hardware implementation of out-of-order processing, where immediately executable instructions can be carried out regardless of the instruction stream sequence. More pipelines will also be used. The P7 will be the first 64-bit 86 family CPU, maintaining object-code compatibility with the 32-bit 86-family CPUs.

(b) Over l00 MB Removable Drives Fight it Out: Magnetic vs Optical

Storage systems with removable disks and capacities of 100 Mbytes or more are beginning to be mounted in personal computers. Within a few years floppy disks with capacities of over 100 Mbytes and optical disks with capacities of 650 Mbytes are expected to be standard storage devices. There types of high-density floppy disks have appeared in the US and Japan, with the Zip drive from Iomega Corp of the US leading in shipments. The LS-120 from Matsushita Industries, Ltd of Japan and the drive being developed by Mitsumi Electric Co, follow closely in second and third place. In optical disk drives, the CD-R drives are receiving high praise because it allows data to be read on a standard CD-ROM drive

三、常用广告用词 (Advertising commentary)

式样齐全 A wide selection of styles
款式入时 Stylish; In up-to-date style
结构合理（结构紧凑）Compact
经久耐用 Durable
安全可靠 Secure , Dependable, Safe
使用方便（携带方便）Handy
备有样本、样品，函索即寄 Catalogues and samples sent upon request
品质优良 Good (Best, Top, Superb, High, Reliable) quality
性能可靠 Reliable (Satisfactory) performance
造型美观 Sleek (Stylish, Attractive) design
外型美观 Luxuriant finish, Attractive design
款式新颖 Styling is in the latest fashion

四、经典广告欣赏

To me，the past is black and white，but the future is always color. (Henness)
对我而言，过去平淡无奇；而未来，却是绚烂缤纷。（轩尼诗酒）

No business too small, no problem too big. (IBM)
没有不做的小生意，没有解决不了的大问题。（IBM）

let's make things better. (Philips)
让我们做得更好。（飞利浦电子）

Connecting People.(Nokia)
科技以人为本。（诺基亚）

Things go better with Coco-Cola. (Coco-Cola)
饮可口可乐，万事如意。（可口可乐）

Make yourself heard. (Ericsson)
理解就是沟通（爱立信）

Time is what you make of it. (Swatch)
天长地久（斯沃奇手表）

Exercises

I. Fill in the blanks in each of the following

1. _____ is an emerging trend that leads to the next step of computing evolution, building on decades of research in virtualization, autonomic computing, grid computing, and utility computing, as well as more recent technologies innetworking, web, and software services.
2. Cloud computing *consists of the* _____ *delivered over the Internet along with the data center hardware and systems software that provide those services.*
3. Cloud computing is *a type of* _____ *consisting of a collection of interconnected andvirtualized computers that are dynamically provisioned and presented as one ormore unified computing resource(s) based on SLAs.*
4. Cloud computing has evolved out of decades of research in different related technologies from which it has inherited some features and functionalities suchas _____, _____, _____, and _____.

II. Answer the following questions

1. What are the characteristics of cloud computing?
2. How many related technologies for cloud computing are discussed by the author and what are these?
3. How many service models of cloud computingare discussed by the author and what are these?
4. Summarizes the business benefits for building software applications using cloud computing technologies.

参考译文

第1章 计算机的历史与未来

1.1 计算机的发明

很难确切地说现代计算机是什么时候发明的。从 20 世纪 30 年代到 40 年代，人们制造了许多类似计算机的机器。但是这些机器大部分不具有当今计算机的所有特征。这些特性是：机器是电子的，具有储存的程序，而且是通用的。

第一个类似计算机的装置是 1941 年由德国的 Konrad Zuse 研制的，叫做 Z3，它是通用型储存程序机器，具有许多电子部件，但是它的存储器是机械的。另一台机电式计算机器是由霍华德·艾坎在 IBM 的资助下于 1943 年在哈佛大学研制的。它被称为自动序列控制计算器 Mark I，或简称哈佛 Mark I。然而，这些机器都不是真正的计算机，因为它们不是完全电子化的。

1.1.1 ENIAC

也许早期最具影响力的类似计算机的装置应该是电子数字积分计算机，或简称 ENIAC。它是由宾夕法尼亚大学的 J. Presper Eckert 和 John Mauchly 研制的。该工程于 1943 年开始，并于 1946 年完成。这台机器极其庞大，重达 30 吨，而且包含 18,000 多个真空管。

ENIAC 是当时重要的成就。它是第一台通用型电子计算机器，并能够执行每秒数千次运算。然而，它是由开关和继电器控制的，必须手工设定。因此，虽然它是一个通用型电子装置，但是它没有储存程序。因此，它没有具备计算机的所有特征。

在 ENIAC 的研制中，一个天才的数学家 Von Neuman（冯·诺伊曼）加入到 Eckert 和 Mauchly 团队，他们一起提出了储存程序计算机的主意。这部机器被称做电子离散变量自动计算机，或简称 EDVAC，是第一部包括了计算机所有特征的机器。然而，直到 1951 年，它一直没有完成。

在 EDVAC 完成之前，其他一些机器建成了，它们吸收了 Eckert、Mauchly 和 Neuman 设计的要素。其中一部是在英国剑桥研制的电子延迟存储自动计算机，或简称 EDSAC，它在 1949 年 5 月首次运行，它可能是世界的第一台电子储存程序、通用型计算机投入运行。在美国运行的第一部计算机是二进制自动计算机，或简称 BINAC，它在 1949 年 8 月投入运行。

1.1.2 UNIVAC I

其他计算机研究先驱一样，Eckert 和 Mauchly 在 1947 年组成了一家公司开发商业计算机。公司名叫 Eckert-Mauchly 计算机公司。他们的目标是设计并建造通用自动计算机或 UNIVAC。因为难以获得财政支持，他们不得不在 1950 年把公司卖给了 Remington Rand 公司。Eckert 和 Mauchly 继续在 Remington Rand 公司从事 UNIVAC 的研制工作，并在 1951 年取得成功。众所周知的 UNIVAC I 机器是世界上第一部商业化计算机。

第一台 UNIVAC I 被交付人口普查局用于 1950 年的人口普查。在投票点关闭后 1 小时之内，第二台 UNIVAC I 被用于预测 D.艾森豪威尔会赢得 1952 年总统大选。UNIVAC I 开始了现代计算机的应用。

1.2 计算机时代

从 UNIVAC I 之后，计算机得到了迅速发展。它们的进化是技术不断变革的结果。这些变革已经创造了四代计算机。

1.2.1 第一代计算机：1951～1958

第一代计算机的特色是使用真空管为其主要电子器件。真空管体积大且发热严重，因此第一代计算机体积庞大，并且需要大量的空调设备保持冷却。此外，因为真空管运行不是很快，这些计算机运行速度相对较慢。

UNIVAC I 是第一代中最早的商业化计算机。如前所述，它在 1951 年被用于人口普查局。它还是第一部应用于商业的计算机。在 1954 年，通用电气接收了 UNIVAC I，并用它进行一些商业数据处理。

然而，UNIVAC I 在第一代计算机中并不是最流行的，这一荣誉属于 IBM 650。IBM 650 在 Remington Rand 造出 UNIVAC I 的后续产品之前的 1955 年首次交付使用。凭借 IBM 650，IBM 占有了大半计算机市场，它在今天仍然保持这一位置。

同时，硬件在进化，软件也在发展。第一部计算机用机器语言编程，但是在第一代计算机期间，程序语言翻译的概念和高级语言出现了。这些主意大部分归功于 Grace Hopper，她在 1954 年是一名海军上尉，学习为哈佛 Mark I 计算机编程。在 1952 年，她开发了第一种编程语言翻译器，在稍后的数年内为其他人所效仿。她还在 1957 年开发了一种称为 Flow-matic 的语言，为 COBOL——今天最广泛应用的商业编程语言奠定了基础。

在第一代计算机期间，其他的软件发展包括 1957 年 FORTRAN 语言的设计。这种语言成为第一种广泛使用的高级语言。同时，第一个简单的操作系统也随着第一代计算机而出现。

1.2.2 第二代计算机：1959～1963

在第二代计算机中，晶体管取代了真空管。第一台全晶体管计算机虽然发明于 1948 年，但直到 1959 年才真正可以使用。晶体管比真空管体积小、价格低，而且运行快、发热少。因此，随着第二代计算机的出现，计算机的体积和成本降低、速度提高，且它们对空调的需要减少。

许多先前不销售计算机的公司随着第二代计算机的出现进入计算机行业，其中在今天仍然制造计算机的公司之一有控制数据公司（DC），他们以制造用于科学工作的高速计算机而闻名。

Remintong Rand，现在叫做 Sperr-Rand 公司，制造了一些第二代 UNIVAC 计算机。然而，IBM 继续称霸计算机行业。最流行的第二代计算机是 IBM 1401，这是一部许多企业使用的中型计算机。

当时所有的计算机都是价值百万元以上的大型计算机。第一台小型计算机产生于 1960 年，价值 12 万美元，它就是由数据设备公司（DEC）制造的 PDP-1。

在这期间软件也在继续发展。许多新的编程语言被发明，包括 1960 年发明的 COBOL。越来越多的企业和组织开始使用计算机以满足他们的数据处理需要。

1.2.3 第三代计算机：1964～1970

作为第三代计算机标志性的技术发展，是在计算机中使用集成电路（简称 IC）。一个集成电路就是包含许多晶体管的一个硅片（芯片）。一个集成电路代替了计算机中的许多晶体管，这使在第二代计算机中的一些趋势得以延续。这些趋势包括计算机体积减小、成本降低、速度提高和对空调的需要减少。

虽然集成电路发明于 1958 年，但是直到 1964 年才出现了第一台广泛使用 IC 的计算机。那一年，IBM 推出了称为 System/360 的大型计算机系列。这一系列的计算机成为使用最广泛的第三代计算机。在 System/360 系列中有许多机型，包括小型的、运行速度相对较慢且价格低廉的机型，到大型的、运行速度非常快且价格昂贵的机型。同时，所有的机型都是兼容的，以便在一个机型上编写的程序可以用于另一个机型。这个在许多计算机系列间兼容的特征被其他第三代计算机制造商所采用。

计算机的第三代也是小型计算机普及的时代。最流行的小型机是由 DEC 制造的 PDP-8。其他公司,包括数据通用公司和惠普(Hewlett-Packard)公司,都在第三代期间开发了小型计算机。

在第三代计算机期间,软件的主要发展是操作系统的复杂化程度提高。第一代和第二代计算机只具有简单的操作系统,许多现代操作系统的特征在第三代期间首先出现。这些特征包括多道程序设计、虚拟存储和分时技术。第一代操作系统主要是批处理系统,但是在第三代期间,交互式系统开始普及,尤其是在小型计算机上。BASIC 语言发明于 1964 年,并由于其交互式特征而在第三代计算机期间大为流行。

1.2.4　第四代计算机:1971～?

第四代计算机比其他三代更难以定义。这一代计算机的特征是一个芯片上包含越来越多的晶体管。首先,出现了一个芯片上具有数百和数千个晶体管的大规模集成电路(LSI),接着出现了一个芯片上具有数万和数十万个晶体管的超大规模集成电路(VLSI)。这个趋势在今天仍在持续。

事实上,并不是每个人都同意第四代计算机这一观点。那些认同的觉得它始于 1971 年,当时 IBM 开发了 System/360 系列计算机的下一系列产品。这些大型计算机称为 System/370,当前的 IBM 计算机虽然不叫做 System/370,但都是从这些计算机直接发展而来的。

小型计算机也在第四代期间迅速增长。最流行的系列是 DEC 公司的 PDP-11 机和 DEC 的 VAX 机,二者在今天的各种机型中仍然有效。

超级计算机首先在第四代中突起。虽然包括 IBM 和 CDC(控制数据公司)在内的许多公司都为科学工作开发了高速计算机,但是直到 1975 年 Cray 研究有限公司推出了 Cray 1,超级计算机才变得有意义。今天,超级计算机是重要的计算机分类。

也许在第四代计算机开始的最重要趋势是微型计算机的增长。随着越来越多的晶体管被集成到硅芯片上,将一整个计算机处理器(称为微处理器)放在一个芯片上终于成为可能。使用微处理器的第一部计算机出现于 1970 年代。第一部专为个人使用设计的微型计算机是 Altair,它于 1975 进入市场。第一部苹果计算机在 1981 年与 IBM 个人计算机一起在市场上销售。今天,微型计算机数目远远超过其他所有类型计算机的总和。

在计算机的第四代期间,软件的发展开始与第三代有所不同。操作系统在逐渐地改进,而新的语言被发明。期间数据库软件被广泛使用。然而,最重要的趋势起因于微型计算机革命。用于微型计算机的软件包随处可得,因此今天大多数的软件可以购得,而不需从头开始开发。

1.2.5　无代计算机

我们可能已经定义了我们最新一代计算机而且开始了计算机的无代时代。即使计算机制造商谈到"第五"和"第六"代计算机,这些说法更多是市场行为,而不是真实的反映。

无代计算机概念的提倡者说,即使科技革新接二连三地迅速出现,也没有一种革新是足够重要到作为另一代计算机的特征。

第 2 章　计算机的基本组成

2.1　介绍

大多数计算机系统,从汽车和日用电器中的嵌入式控制器到个人计算机和大型主机,都具有相同的基本组成。其基本组成包括三个主要部件:CPU、存储器子系统和 I/O 子系统。这些部件的一般组成如图 2-1 所示。

本章我们首先讲述计算机系统中用来连接计算机各部件的系统总线。然后再来考察指令周期,以及计算机在读取、解码和执行一条指令时所发生的操作顺序。

2.2 系统总线

从物理上来说，总线就是一组导线。计算机的部件就是连在总线上的。为了将信息从一个部件传到另一个部件，源部件先将数据输出到总线上，然后目标部件再从总线上接受这些数据。随着计算机系统复杂性的不断增长，使用总线要比在每个设备之间使用直接连接有效得多（就减少连接数量而言）。与大量的直接连接相比，总线使用较少的电路板空间，耗能更少，并且在芯片或组成 CPU 的芯片组上需要较少的引脚。

图 2-1 所示的系统包括三组总线。最上面的是地址总线。当 CPU 从存储器读取数据、指令或写数据到存储器时，它必须指明将要访问的存储器单元地址。CPU 将地址输出到地址总线上，然后存储器从地址总线上读取地址，并且用它来访问正确的存储单元。每个 I/O 设备，比如键盘、显示器或者磁盘，同样都有一个唯一的地址。当访问某个 I/O 设备时，CPU 将此设备的地址放到地址总线上，每一个设备均从总线上读取地址并且判断自己是否就是 CPU 正要访问的设备。与其他总线不同，地址总线总是从 CPU 上接收信息，而 CPU 从不读取地址总线。

数据是通过数据总线传送的。当 CPU 从存储器中读取数据时，它首先把存储器地址输出到地址总线上，然后存储器将数据输出到数据总线上，这样 CPU 就可以从数据总线上读取数据了。当 CPU 向存储器中写数据时，它首先将地址输出到地址总线上，然后把数据输出到数据总线上，这样存储器就可以从数据总线上读取数据并将它存储到正确的单元中。对 I/O 设备读写数据的过程与此类似。

控制总线与以上两种总线都不相同。地址总线由 n 根线构成，n 根线联合传送一个 n 位的地址值。类似地，数据总线的各条线合起来传输一个单独的多位值。相反，控制总线是单根控制信号的集合。这些信号用来指示数据是要读入 CPU 还是要从 CPU 写出，CPU 是要访问存储器还是要访问 I/O 设备，是 I/O 设备还是存储器已就绪要传送数据等等。虽然图 2-1 所示的控制总线看起来是双向的，但它实际上（主要）是单向（大多数都是）信号的集合。大多数信号是从 CPU 输出到存储器与 I/O 子系统的，只有少数是从这些子系统输出到 CPU 的。在介绍指令周期和子系统接口时，我们将详细地讨论这些信号。

一个系统可能具有分层次的总线。例如，它可能使用地址、数据和控制总线来访问存储器和 I/O 控制器。I/O 控制器可能依次使用第二级总线来访问所有的 I/O 设备，第二级总线通常称为 I/O 总线或者局部总线。

2.3 指令周期

指令周期是微处理器完成一条指令处理的步骤。首先，微处理器从存储器读取指令，然后将指令译码，辩明它取的是哪一条指令。最后，它完成必要的操作来执行指令（有人认为在指令周期中还要包括一个附加的步骤来存储结果，这里我们把该操作当作执行功能的一部分）。每一个功能——读取、译码和执行都包括一个或多个操作。

我们从微处理器的存储器中取指令开始讲述。首先，微处理器把指令的地址放到地址总线上，然后，存储器子系统从总线上输入该地址并予以译码，去访问指定的存储单元。（译码是如何进行的，我们将在后面介绍存储器子系统进行更为详细的讨论。）

当微处理器为存储器留出充足的时间来进行地址译码和访问所需的存储单元之后，微处理器发出一个读（READ）控制信号。当微处理器准备好可以从存储器或是 I/O 设备读数据时，它就在控制总线上发一个读信号。（一些处理器对于这个信号有不同的名字，但所有处理器都有这样的信号来执行这个功能。）根据微处理器的不同，读信号可能是高电平有效（信号=1），也可能是低电平有效（信号=0）。

读信号发出后，存储器子系统就把要取的指令码放到计算机的数据总线上，微处理器就从数据总线上输入该数据并且将它存储在其内部的某个寄存器中。至此，微处理器已经取得了指令。

接下来，微处理器对这条指令译码。每一条指令可能要有不同的操作序列来执行。当微处理器对该指令译码时，它将确定处理的是哪一条指令，以便选择正确的操作序列去执行。这一步完全在微处理器内完成，不需要使用系统总线。

最后介绍微处理器执行该指令。指令不同，执行的操作序列也不同。执行过程可以是从存储器读取数据，写数据到存储器，读或写数据到 I/O 设备，执行 CPU 内部操作或者执行多个上述操作的组合。下面我们从系统的角度来看计算机是怎样执行这些操作的。

微处理器从存储器读取数据所执行的操作序列，同从存储器中去一条指令是一样的。毕竟取指令就是简单地从存储器中读取它。图 2-2（a）显示了从存储器中读取数据的操作时序。

在图 2-2 中，注意最上面的符号 CLK，它是计算机的系统时钟，微处理器用系统时钟使其操作同步。在一个时钟周期（系统时钟的 0/1 序列）的开始位置，微处理器将地址放到总线上。一个时钟周期（允许存储器对地址译码和访问数据的时间）之后，微处理器才发出读信号。这使得存储器将数据放到数据总线上。在这个时钟周期之内，微处理器从系统总线上读取数据，并存储到它的某个寄存器中。在这个时钟周期结束时，微处理器撤消地址总线上的地址，并撤消读信号。然后存储器从数据总线上撤消数据，也就完成了存储器的读操作。

存储器写操作的时序如图 2-2（b）所示。在第一个时钟周期，处理器将地址和数据放到总线上，然后在第二个时钟周期开始时发出一个写（WRITE）控制信号（或与之等价的信号）。像读信号促使存储器读取数据一样，写信号促使存储器存储数据。在这个时钟周期的某个时刻，存储器将数据总线上的数据写入地址总线指示的存储单元内。当这个时钟周期结束，微处理器从系统总线上撤消地址、数据及写信号后，就完成了存储器的写操作。

I/O 的读写操作与存储器的读写操作类似。处理器可以使用存储器影射 I/O 或者是单独 I/O。如果处理器支持存储器影射 I/O，则它遵循从存储器读写数据同样的操作顺序，该顺序如图 2-2 所示（记住，在存储器影射 I/O 中，处理器把一个 I/O 端口当作某个存储单元，当然 I/O 的数据访问同存储器的数据访问一样的）。使用单独 I/O 的处理器遵循同样的处理过程，但是另有一个控制信号用以区别是 I/O 访问还是存储器访问（使用单独 I/O 的 CPU 允许一个存储单元和某个 I/O 端口具有相同的地址，因此需要这一额外的信号加以区分）。

最后，考虑一下完全在微处理器内部执行的指令。相对简单 CPU 的 INAC 指令和 8085 的 MOV r1, r2 指令的执行都不要访问存储器和 I/O 设备。按照指令译码的结果，这些指令的执行不会用到系统总线。

2.4 CPU 的组成

CPU 控制整个计算机。它从存储器中取指令，提供存储器需要的地址和控制信号。CPU 对指令译码并且控制整个执行过程。它执行一些内部操作，并且为存储器和 I/O 设备执行指令提供必要的地址、数据和控制信号。除非 CPU 激发，否则，计算机什么事情都不会发生。

CPU 内部有三大分区，如图 2-3 所示。寄存器区，顾名思义，它包括一组寄存器、一条总线或其他通信机制。微处理器指令集结构中的寄存器就属于 CPU 的这一分区。系统的地址和数据总线与寄存器交互。此分区还包括程序员不能直接访问的一些寄存器。相对简单的 CPU 含有寄存器用以锁存正在访问的存储器地址，还有暂存器以及指令集结构中没有的其他寄存器等。

在指令周期的取指阶段，处理器首先将指令的地址输出到地址总线上。处理器有一个寄存器叫做程序计数器，CPU 将下一条要取的指令地址存放在程序计数器中。在 CPU 将地址输出到系统的地址总线之前，必须从程序计数器中取出该地址。在指令结束前，CPU 从系统总线上读取指令码，它把该指令码存储在某个内部寄存器中，该寄存器通常称作指令寄存器或其他相似的名字。

算术逻辑单元执行大部分的算术逻辑运算，如加法、逻辑与等运算。它从 CPU 的寄存器取得操作数，然后将运算结果再存回到寄存器区。由于必须在一个时钟周期内完成操作，因此 ALU 只采用组合逻辑构造而成。相对简单的 CPU 和 8085 微处理器中的 ADD 指令在执行中都有使用 ALU。

同 CPU 控制整个计算机（除了其他功能外）一样，控制单元控制着 CPU。这个单元产生内部控制信号，促使寄存器装载数据，自动加 1 或清零，输出它的内容，使得 ALU 完成正确的操作等等。这些信号作为控制信号显示在图 2-3 中。控制单元从寄存器区取得一些数据用以产生控制信号，这些数据包括指令码和某些标志寄存器的值。控制单元也产生系统控制总线上的信号，例如 READ，WRITE，$IO/\overline{M}$ 信号等。典型的一个微处理器执行取指令、译指令和执行指令等一系列的操作。通过以正确的顺序激发这些内部或外部控制信号，控制单元使 CPU 和计算机的其余部分完成正确处理指令所需要的操作。

以上对 CPU 的描述并不完整。现在的处理器拥有更加复杂的特征以提高其性能。这些机制中有一种是指令流水线技术，它允许 CPU 在执行一条指令的同时取出另一条指令。

第 3 章　二进制和布尔代数

用于计数的十进制被现代文明广泛采用，以至于我们很少考虑其他数制的使用可能性。但是，如果期望使用十个手指可以表示的十进制方法作为机器结构最有效的数制是不合理的。事实上，很少使用，但却是很简单的二进制数为机器提供了一个最自然和有效的数制。

3.1　十进制

我们当前的数字系统有 0、1、2、3....9 十个单独的符号，称之为阿拉伯数字。如果不使用位置符号，我们只能数到 9 就被迫停下来，或发明更多的符号。在罗马数字里可以找到早期符号类型的例子，它们基本上采用加法制：III＝I＋I＋I，XXV=X+X+V，当数值增加时采用新符号（X、C、M 等）。这样 V 就不是 IIIII＝5。罗马数字中位的唯一重要性在于这个符号处于另一个符号之前或之后（Ⅳ=4，Ⅵ=6）。如果你要用 XIV 乘 XII，很容易看出这个数字系统是笨拙的。用罗马数字计算太难了，以至于早期的数字家几乎完全被迫要先在算盘或演算板完成算术运算，然后再把结果翻译成罗马数字形式。在这样的数字系统中，纸和笔运算达到以难置信的复杂和困难。事实上，在早期文明中能进行这样的加法和乘法运算被看作是一项伟大的成就。

现在可以看到我们的数字系统简单明了的巨大优势，想要说出任意想到的数字，只需要学会基本数字和进位符号，再记住加法和乘法表及学会一些简单规则，就可能完成所有的算术运算。看一下用现在数制计算 12×14 的简单性。

如果我们注意到说 '一百六十八' 时，数字 168 的实际意义就能更清楚地看出来。基本上，这个数字是（1×100）+6×10）+8 的紧缩形式。更重要的是每个数字的值由它的位置来决定。例如 2000 中的 2 和 20 中的 2 的值是不同的。口头表达这些为 '二千' 和 '二十'。从 10 到 20 我们发明出不同的口头表示方式，但是从 20 起往上，我们只在 10 的权位上断开。书写出的数字总是紧凑的，不论写出的整数大小，只用 10 个基本数字。十进制使用进位符号表示数字的通则是

$$\alpha_{n-1}10^{n-1} + \alpha_{n-2}10^{10-2} + ... + \alpha_0 10^0$$

不同位上的整数用 a_{n-1}，a_{n-2}，…a_0 表示，n 表示十进制小数点左面数字的数量。

基数是定义在数字系统中每一位上的不同数字。十进制数有一个 10 的基数，这表示它有 10 个不同数字。（0、1、2....9），其中的任意一个都可以用在数字的每个位置上。历史上记录了几种使用过的其他数制：五进

制有 5 个数字作为基数，在爱斯基摩人和北美印第安人中流行；十二进制（12 个基数）可以在钟表、英尺、英寸以及以打记数中看到。

3.2 二进制

一位十七世纪的德国数学家 Gottfrid Wilhelm von leibniz 提倡二进制，只使用 0 和 1 两个符号作为基数。一位杰出的数学家提倡用如此简单的数字似乎有些奇怪，因为要注意到的是他还是一位杰出的哲学家。Leibniz 提倡二进制的原因似乎有些神秘。他感到用零代表虚无和一代表上帝之间的相似之处很有优势。

不管 leibniz 提出的理由有多好，二进制在过去的十年里变得很流行。现在的数字计算机以二进制原理构成，当前的迹象表明将来的机器将继续使用这些系统操作。

早期计算机的基本元件是继电器和开关，开关或操作继电器可以看作是基本的二进制性质。那就是开（1）或关（0）两种状态。更先进一些的计算机主要电器元件是类似于收音机和电视机的晶体管。使设计者采用这些装置是因为其具有良好的可靠性，它们基本上处于两个状态之一，完全导通或截止。在这种电路和电灯之间可以做简单的模拟。任意给定的时间里，电灯（或晶体管）处于导通或截止状态之一。例如即使电灯泡很旧了，仍可以轻易区分开它是开或关的状态。同样的事情也可以在收音机中看到，当收音机老化了，音量就减低了，我们用调高音量来补偿。所以即使当收音机很旧了，还能容易区分是开或关状态。

由于计算机中大量采用了电子器件，强烈要求利用它们的一些特性，即当特性稍有变化时不至影响性能。做到这些的最好方法是采用双稳态电路。

3.2 布尔代数

布尔代数的概念最初是由英国数学家 George Boole 于 1847 年提出来的，从那时起，代数学家和逻辑学家们开始广泛地发展 Boole 最初的概念，并使之更加精练。由于布尔代数、集合代数、逻辑学和二进制算术之间的内在联系，使得布尔代数的理论在电子计算机的发展中起到举足轻重的作用。

布尔代数最直觉的发展产生于集合代数的概念。设 S={a,b,c} 和 T={a,b,c,d,e} 分别为两个含有三个和五个元素的集合。由于 S 中的每一个元素（a,b,c）都属于 T，所以我们说 S 是 T 的一个子集。由于 T 有五个元素，因而 T 共有 2^5 个子集，这是因为我们可以选择任何一个元素使其包含于某个子集中或从该子集中删除。应该注意到这 32 个子集中包含 T 本身和空集（空集即不含任何元素的集合）。如果 T 包含了所讨论的所有元素，则称之为全集。给定 T 的一个子集，例如子集 S，我们可以定义一个关于全集 T 的 S 的补集。其中正好包含那些不在子集 S 中而在 T 中的元素。

于是，如上定义的集合 S 就有一个它的补集（相对于集合 T），$\bar{S}$={d,e}。任何两个集合（已给定集合的若干子集）的并集包含了出现于这两个子集中某一个集合或同时出现于这两个集合中的所有元素；两个集合的交集包含了同时出现于这两个集合中的元素。我们用符号"∪"来表示两个集合的"并（运算）"，用"∩"来表示两个集合的"交（运算）"。例如，如果 B={b,d,e},那么，B∪S={a,b,c,d,e}, B∩S={b}。

虽然我们可以定义其他一些集合运算，但求补、并和交运算是我们最感兴趣的三个集合运算。一个布尔代数就是一个有限集或无限集，以及建立在该有限集或无限集上的三种运算——否定、加或乘，这三个运算分别对应于集合的求补、并和交运算。在布尔代数的元素中有两个特殊的元素：0, 对应于空集；1, 对应于全集。对于一个布尔代数中任意给定元素 a，都有一个唯一的补 a'，它满足 a+a'=1 和 aa'=0。布尔加和布尔乘与普通的加和乘一样，满足结合律和交换律，但除此之外含有一些不太相同的特性。其主要特性由表 3-2 给出，其中 a, b 和 c 是一个布尔代数中的任意元素。

表 3-2

分配律	a(b+c)=ab+ac
	a+(bc)=(a+b)(a+c)
同一律	a+a=a
	aa=a
吸收律	a+ab=a
	a(a+b)=a
德·摩根定理	(a+b)'=a'b'
	(ab)'=a'+b'

由于 n 个元素的有限集有且只有 2^n 个子集，而且很显然有限布尔代数一定是有限集合代数，所以对某个整数 n 而言，每个有限布尔代数也有且只有 2^n 个元素。例如，上文定义的集合 T 的集合代数就对应一个有 32 个元素的布尔代数。

虽然我们可以用不同的符号来表示布尔代数中的每一个元素，但最常用的方法是用一个有 n 个分量的二进制向量来表示一个有限布尔代数的 2^n 个元素。用这样一种表示方法，布尔代数的所有运算都以分量的形式完成，而每一个分量被认为是一个独立的二值布尔代数。这种做法对应于用二进制向量来表示一个有限集的子集。例如，由于集合 T 有 5 个元素，所以我们可以用 5 个分量的二进制向量表示它的子集，其中每一个分量表示集合 T 的一个元素。向量中的第 i 个分量为数字 1 表示集合 T 的第 i 个元素在某一特定子集中，用数字 0 表示不在某一特定子集中。于是，子集 S={a,b,c} 可用二进制向量表示为 {1, 1, 1, 0, 0}。集合运算变成了向量分量上的布尔运算。集合的这种表示方法以及相应的布尔或逻辑运算，对于信息检索是非常有用的。由于这一原因，文件的集合和查询特性可以很容易而迅速地得到匹配。

第 4 章 基本数据结构

4.1 栈和队列

栈和队列都是动态的集合，其中的元素由预先定义的删除操作可将其去除。在栈中，最近被插入的元素最先被删除，即栈是按后进先出（LIFO）的原则进行的；类似的，在队列里，被删除的元素是最早进入队列的，即队列是以先进先出（FIFO）的原则进行的。在计算机上，对于栈和队列有几种有效的实现方法，这一节我们给出如何用简单的数组来实现它们。

4.1.1 栈

栈中插入元素的操作通常称为入栈（PUSH），删除操作称为出栈（POP），这里不考虑删除的那个元素。这些名字来源于实际的堆栈，就像餐厅里使用的带有弹簧座的一叠盘子：因为只有顶部的盘子是方便取用，所以盘子从这一叠中取出的顺序与其被放入的顺序是相反的。

如图 4-1 所示，我们可以定义一个大小为 n 的栈，由数组 S[1, n] 来实现。数组用一个栈顶指针 *top[S]* 来指示最近被插入的元素；栈中的元素个数为 S[1, *top[S]*]，其中 S[1] 是栈底元素，S[*top[S]*] 是栈顶元素。

图 4-1：顺序栈的实现。栈中元素处于浅色阴影的位置。*(a)* 栈 S 有 4 个元素，栈顶元素是 9；*(b)* 栈 S 执行 PUSH（S, 17）、PUSH（S, 3）操作后；*(c)* 栈 S 执行了 POP（S）操作后，使最近入栈的元素 3 出栈。虽然

元素 3 还在数组中，但是它已不再位于栈内，当前的栈顶元素是 17。

$top[S] = 0$ 表示栈中没有元素，是空栈。通过查询操作 STACK-EMPTY 可判断栈是否为空。如果对空栈执行出栈操作，我们称之为下溢，通常是一个错误。如果栈顶指针 $top[S]$ 超出 n，那么栈上溢。（在伪码实现时，我们不考虑栈上溢的情况。）

每个对栈所进行的操作都可以由几行代码实现：

STACK-EMPTY(S)

1 **if** $top[S] = 0$

2 **then return** TRUE

3 **else return** FALSE

PUSH(S, x)

1 $top[S] \leftarrow top[S] + 1$

2 $S[top[S]] \leftarrow x$

POP(S)

1 **if** STACK-EMPTY(S)

2 **then error** "underflow"

3 **else** $top[S] \leftarrow top[S] - 1$

4 **return** $S[top[S] + 1]$

图 4-1 给出执行入栈和出栈这些修改操作后的结果，这三个操作的时间复杂度都是 $O(1)$。

4.1.2 队列

我们称对队列进行的插入操作为入队（ENQUEUE），删除操作为出队（DEQUEUE）；同栈中的出栈操作一样，出队操作不考虑其中的元素。队列的先进先出原则就像排队等待注册的一行人群。队列有一个头部和一个尾部。当一个元素入队，它就成为队列的尾部，就像一个新来的学生成为队尾一样；出队的元素总是队列头部的元素，像是站在队列最前的学生是排队最久的人一样。（庆幸的是，对于计算机元素，我们不必担心插队的情况。）

图 4-2 给出了用数组 $Q[1 \_ n]$ 来实现队列的一种方法，其中队列最多可有 $n-1$ 个元素。队列有一个头指针 $head[Q]$，用来指出队列的头部；尾指针 $tail[Q]$ 指向下一个将要入队的元素所在位置。队列中元素的所在位置就是 $head[Q], head[Q]+1,..., tail[Q]-1$,其中结点 1 紧跟在结点 n 后。$head[Q] = tail[Q]$ 表示队列为空，初始状态为 $head[Q] = tail[Q] = 1$。

当队列为空时，执行出队操作将导致队列的下溢；当 $head[Q] = tail[Q] + 1$ 时，即队列为满，执行入队操作会引起队列的上溢。

图 4-2：由数组 $Q[1 \_ 12]$ 实现的队列。队列中元素处于浅色阴影的位置。*(a)* 队列有 5 个元素，位于 $Q[7 \_ 11]$；*(b)* 队列执行了操作 ENQUEUE（Q, 17）、ENQUEUE（Q, 3）和 ENQUEUE（Q, 5）后；*(c)* 队列执行了操作 DEQUEUE（Q）后，使原头部元素 15 出队。这时的头部元素值为 6。

在操作 ENQUEUE 和 DEQUEUE 中，我们不考虑对上溢和下溢的错误检测。

ENQUEUE(Q, x)

1 $Q[tail[Q]] \leftarrow x$

2 **if** $tail[Q] = length[Q]$

3 **then** $tail[Q] \leftarrow 1$

4 **else** $tail[Q] \leftarrow tail[Q] + 1$

DEQUEUE(Q)

1 $x \leftarrow Q[head[Q]]$
2 **if** $head[Q] = length[Q]$
3 **then** $head[Q] \leftarrow 1$
4 **else** $head[Q] \leftarrow head[Q] + 1$
5 **return** x

图 4-2 给出执行入队和出队操作后的结果，每个操作的时间复杂度都是 $O(1)$。

4.2 链表

链表这个数据结构中的对象都以线性顺序排列的，但不同于数组的顺序是由数组下标确定，链表的顺序是由每个对象的指针确定。链表为动态集合提供了一个简单、灵活的表示形式。

如图 4-3 所示，双向链表 L 中的每个元素都是一个实体，由一个数值域和两个指针域组成，其中两个指针分别是：*next* 和 *prev*，实体也可以包含子数据。给定表中一个元素 x，*next*[x]指出它在链表中的后继，*prev*[x]指出它的前趋。*prev*[x] = NIL 表示元素 x 没有前趋，因此 x 是第一个元素，即队头元素；如果 *next*[x] = NIL，表示 x 没有后继，也就是最后一个元素，队尾元素。指针 *head*[L]指向队列中的第一个元素，如果 *head*[L] = NIL 表示队列为空。

图 4-3：*(a)*一个双向链表 L 描述了一个含有{1, 4, 9, 16}等元素的动态集合。表中的每一个元素都是一个包含有值域和指针域（如箭头所示）的实体，指针分别指向前一个和下一个实体。尾部元素的 *next* 指针域和头部元素的 *prev* 指针域都为空（NIL），由一个斜线号表示。指针 *head*[L]指向队列的头部；*(b)*执行 LIST-INSERT（L, x）操作后，其中 *key*[x] = 25，链表的头部元素是值为 25 的新对象，这个新对象指向原头部元素，其值为 9；*(c)*执行 LIST-DELETE（L, x）操作后的结果，x 指向的对象值为 4。

一个链表可以是下面几种形式之一：单向链表或双向链表，有序的或无序的，循环的或非循环的。对于单向链表，我们就不必对每个元素使用 prev 指针了。如果一个表是有序的，表的线序和表中存储的元素值的线序是一致的，即：最小值的元素是队列的头部元素，最大值的元素是队尾元素。如果一个表是无序的，其中的元素可以以任何顺序出现。

在一个循环表中，队头的 prev 指针指向队尾元素，队尾的 next 指针指向队头的元素，整个链表的元素组成一个环。在这一节的以后内容中，我们假定，我们所讨论的都是无序的双向列表。

4.2.1 链表的查找

通过一个简单的线性查询，操作 LIST-SEARCH（L, k）找出表 L 中第一个值为 k 的元素，并返回它的指针。如果表中没有值为 k 的对象，那就返回空值（NIL）。如图 4-3(a)中的链表，操作 LIST-SEARCH（L, 4）返回第三个元素的指针，而 LIST-SEARCH（L, 7）返回空值。

LIST-SEARCH(L, k)

1 $x \leftarrow head[L]$
2 **while** $x \neq$ NIL and $key[x] \neq k$
3 **do** $x \leftarrow next[x]$
4 **return** x

查找有 n 个对象的列表，操作 LIST-SEARCH 在最坏情况下的时间复杂度为 $O(n)$，因为可能需要查找整个列表。

4.2.2 链表的插入

给定一个元素 x，它的值域已经设定，操作 LIST-INSERT 将 x 插入链表的头部,如图 4-3(b)所示。

LIST-INSERT(*L*, *x*)
1 *next*[*x*] ← *head*[*L*]
2 **if** *head*[*L*] ≠ NIL
3   **then** *prev*[*head*[*L*]] ← *x*
4 *head*[*L*] ← *x*
5 *prev*[*x*] ← NIL

操作 LIST-INSERT 在一个 n 元列表上的运行时间是 $O(1)$。

4.2.3 链表的删除

执行 LIST-DELETE 操作,可将元素 x 从链表 L 中去除。必须先给定指向 x 的指针,通过更新指针将 x 从链表中删除。如果需要删除给定值的元素,我们必须先执行 LIST-SEARCH 操作,检索指向元素的指针。

LIST-DELETE(*L*, *x*)
1 **if** *prev*[*x*] ≠ NIL
2   **then** *next*[*prev*[*x*]] ← *next*[*x*]
3   **else** *head*[*L*] ← *next*[*x*]
4 **if** *next*[*x*] ≠ NIL
5   **then** *prev*[*next*[*x*]] ← *prev*[*x*]

图 4-3(c)显示了链表中如何删除一个元素。LIST-DELETE 的运行时间是 $O(1)$,但如果我们想删除一个指定值的元素,最坏情况下的时间复杂度为 $O(n)$,因为需要先执行 LIST-SEARCH 操作。

4.2.4 标志

如果我们不考虑链表头部和尾部的边界情况,对操作 LIST-DELETE 的编码会更简单。

LIST-DELET′(*L*, *x*)
1 *next*[*prev*[*x*]] ← *next*[*x*]
2 *prev*[*next*[*x*]] ← *prev*[*x*]

利用标志这样的哑元实体,我们可以简化边界条件的判定。例如,假设我们为链表 L 设置一个实体 *nil*[*L*],它的值域为空,但是它却指向链表中所有其他的元素。我们在代码行中,所使用到 NIL 之处,都可以由标志 *nil*[*L*]代替。如图 4-4 所示,这将一个普通的双向链表变成一个带标志的双向链表,其中标志于表头和表尾之间,即:指针域 *next*[*nil*[*L*]]指向表头,*prev*[*nil*[*L*]]指向表尾。类似地,队尾的 next 域和队头的 prev 域都指向 *nil*[*L*]。由于 *next*[*nil*[*L*]]指向表头,我们可以完全去掉指针 *head*[*L*],取而代之是使用 *next*[*nil*[*L*]]。一个空表只包含有标志,因为指针域 *next*[*nil*[*L*]]和 *prev*[*nil*[*L*]]都指向 *nil*[*L*]。

图 4-4:一个带标志的循环双向链表。标志 *nil*[*L*]位于头部和尾部之间。因为我们可以通过 *next*[*nil*[*L*]]来访问链表的头部,所以指针 head[L]不再需要。*(a)*一个空的链表;*(b)*图 4-3(a)中的链表,头部元素的数值是 9,尾部元素的数值是 1;*(c)*执行操作 LIST-INSER′(*L*, *x*),其中 *key*[*x*] = 25,新对象成为链表的头部;*(d)*删除数值为 1 的对象后的链表,新的队尾元素的数值是 4。

操作 LIST-SEARCH 的编码与之前的保持一致,但是像上面指出的,NIL 和 *head*[*L*]作了修改。

LIST-SEARC′(*L*, *k*)
1 *x* ← *next*[*nil*[*L*]]
2 **while** *x* ≠ *nil*[*L*] and *key*[*x*] ≠ *k*
3   **do** *x* ← *next*[*x*]
4 **return** *x*

我们用两行的操作 LIST-DELET'来删除链表中的一个元素，我们用下面的操作实现在链表中插入一个元素。
LIST-INSER' (L, x)

1 $next[x] \leftarrow next[nil[L]]$
2 $prev[next[nil[L]]] \leftarrow x$
3 $next[nil[L]] \leftarrow x$
4 $prev[x] \leftarrow nil[L]$

图 4-4 给出在一个具体链表上执行 LIST-INSER' 和 LIST-DELET'后的结果。标志几乎不能在数据结构操作中,线性地减少时间,但是可以减少常数因子的时间。比如,使用标志简化了对链表的编码,但在 LIST-INSER' 和 LIST-DELET'操作中, 节省的时间仅是 $O(1)$；而在另一种情况中, 使用标志减少了循环中的代码, 也就降低了运行时间中 n 或 n2 的系数。

标志并不是可以任意使用的, 如果对许多小的链表使用标志而造成的额外存储将浪费巨大的存储空间。

第 5 章　 操作系统

5.1　 操作系统的功能

操作系统的功能包括：
1. 资源分配及相关功能。
2. 用户接口功能。

资源分配功能负责实现计算机系统的用户共享资源。它主要将资源和提出需求的程序绑定在一起, 即使资源和程序相关联。与之相关的一些功能包括实现不同用户在共享资源时的保护, 以免发生相互干扰。用户接口功能帮助用户创立并使用适当的计算结构, 这一功能通常涉及命令语言或菜单的应用。

5.1.1　 资源分配及相关功能

资源分配功能是分配资源供用户计算使用。资源可分为系统提供的资源（如 CPU、存储器区域及 I/O 设备群) 和用户创建的资源（如由操作系统管理的文件等）。资源分配的标准根据资源的分类确定, 系统资源的分配要考虑资源利用的效率, 而用户创立资源的分配则基于该资源的创立者所设定的特种限制, 比如访问权限。

资源分配通常采取以下两种策略：
1. 资源分区。
2. 从资源池中分配。

在资源分区方式中, 操作系统预先决定把哪些资源分配给某个用户计算使用, 这种方法也称为静态分配, 因为分配是在程序执行前进行的。静态资源分配易于实现, 但由于它不是从程序的实际需要出发, 而是根据程序预先提出的需求来做决定, 所以容易导致系统利用率下降。在后一种分配方式中, 操作系统维护一个公共资源池, 并按照程序的需要对资源进行分配。这样, OS 只在程序提出对一个资源的需求时才进行资源分配, 因为分配是在程序执行的过程中进行的, 这种方式也称为动态分配。动态存储分配的资源利用率较高, 因为它是在程序需要资源时才进行分配。

操作系统可以利用资源表作为资源分配的中心数据结构。表中包含系统的每一资源单位项数据项中记录资源单位的名称或地址以及当前状态, 即它是空闲的还是已经分配给某一程序。当程序对某一资源提出请求后, 若该资源是空闲的, 则它将被分配给那个程序。若系统中同一资源类中存在许多资源单位, 程序的资源请求只指明要求哪类资源而由操作系统检查该类中是否有可用的资源单位可以分配。

在资源分区分配中，操作系统依据系统中的资源数目和程序数目决定如何进行资源分配。例如，操作系统可能会决定一个程序可以使用 1MB 内存、2000 个磁盘块及 1 个监视器，这一系列资源可以作为一个分区。系统可以提前定义出一系列这样的区，资源表中为每一资源区保存相应的数据项。当新的程序开始时将分配给它一个可用的分区。

资源抢占

资源被一系列程序分享可以有多种不同的方式，如：

- 串行共享
- 并行共享

在串行共享中，一个资源为一个程序独占，当资源被释放后，它在资源表中被标记为"空闲"，这时它才能被分配给下一个程序。在并行共享中，两个以上的程序可以同时使用同样的资源。并行共享的例子是数据文件，其他资源通常不能被并行共享。在本文中以后若无特殊说明，所有资源都被假设为只能串行共享。

当资源可以串行共享时，系统可以在程序明确地提出释放分配的请求后，对资源进行释放。或者，系统也可强行释放资源，这称为资源抢占。

操作系统依靠对系统资源的抢占确保程序对资源的平等利用，或实现一些系统级的目标。对于资源被抢占的程序来说，除非把被抢占的资源单位或同一资源类的其他资源单位重新分配给它，否则无法继续执行。我们称抢占 CPU 为抢占，而称抢占其他资源为资源抢占。

CPU 共享

CPU 只能串行共享，因此它一次只能执行一个程序，其他程序必须依次等待。通常情况下，系统会要求平等对待所有的程序，用抢占来释放 CPU 以执行其他程序。因此决定该执行哪个程序并执行多长时间是一个十分重要的功能，这一功能被称为 CPU 调度或简称调度。显然，资源分区不适用于 CPU 共享，同此，从资源池中分配成为惟一的选择。

存储器共享

和 CPU 一样，存储器也不能并行共享。但与 CPU 不同的是，可以把存储器的不同部份看作不同的资源，因此可以增加它的可用性。资源分区和基于资源池的分配方式都适用于存储器资源管理。存储器抢占也可用于提高程序对存储器的可用性，但由于对不同的存储器抢占技术都有各自的专用术语，故在我们的讨论中很少使用"存储器抢占"的说法。

5.1.2 用户接口相关功能

用户接口的目的是为用户提供使用操作系统资源（主要是 CPU）的接口来完成用户的计算请求。操作系统用户接口一般指命令语言，用户通过命令建方合适的计算结构以满足计算要求。

不同的计算结构可由操作系统决定。计算结构举例如下：

1. 单一程序。
2. 单一程序的序列。
3. 程序集合。

以上这些计算结构将在以后各章中定义和描述，此处只指出一些简单的特征。例如每个程序都是由用户通过用户接口分别进行初始化的，单一程序是由在给定的数据集上的一个程序执行组成的，用户通过命令启动程序。程序可以分为两种：顺序的和并发的。顺序程序符合传统的程序概念，是最简单的计算结构。而在并发程序中，程序的不同部分可被同时执行。为此，操作系统需要了解这些可以同时执行的不同部分的实体，操作系统的用户接口通常不支持这一功能。在本章中，假设所有的程序都是顺序执行的。

在一个程序序列中，每个程序是由用户单独启动的，但程序之间并非相互独立，只有当前一程序执行成功后，下一个程序的执行才有意义，但由于程序是独立启动的，它们之间相互的接口由用户明确设定。在程序集合中，用户在其命令中为集合中涉及的程序命名。由此它们将通过用户接口被操作系统识别，程序间的接口由操作系统处理。

5.2 FreeBSD，Linux 和 Windows 2000

可靠性

FreeBSD 系统非常健壮，有众多的常年稳定运行的服务器可以证明这一点。新的软件升级使文件系统优化了磁盘 I/O 以获得更高的性能，同时也保证了像数据库这样的基于事务的应用软件更加稳定运行。

Linux 系统也是以稳定而著称，安装 Linux 系统的服务器可以长年稳定工作。但是 Linux 系统的磁盘 I/O 是非同步的，这降低了基于事务的操作的稳定性，在系统出错或者电源失效的情况下会导致文件系统的错误，但是对一般的用户而言，Linux 系统已经非常可靠了。

所有的 Windows 用户都习惯了蓝屏错误，可靠性差是 Windows 系统的主要缺陷。Windows 2000 系统修正了部分主要缺陷，但是由于系统代码非常庞大，导致存在很多的稳定性隐患。Windows 2000 系统使用了大量的系统资源，而且系统很难持续运行几个月，原因是由于文件系统产生了大量的碎片和内存分配的混乱。

性能

FreeBSD 系统非常适用于高性能的网络应用，FreeBSD 系统在同等硬件环境下要比其他的系统运行的好。Internet 上最大和最忙的公众服务器都采用了 FreeBSD 系统，在 ftp://ftp.cdrom.com/网站，采用 FreeBSD 系统支撑了每天 1.2TB 的下载量。该系统在 Yahoo!，Qwest 以及其他著名网站，都作为其主要服务器采用的操作系统，并且能够高效稳定的处理高负载的网络流量。

Linux 系统对大多数应用来说性能良好，但是在网络负载很重的情况下表现不佳。在同样的硬件环境下，Linux 的网络性能要比 FreeBSD 系统低 20%-30%。Linux 对此问题已有所改进，在 2.4 版本内核中，它将采用一种新的虚拟内存技术，这种技术类似于 FreeBSD VM。由于 Linux 和 FreeBSD 都是开放源代码，因此好的技术可以在两种系统中共享，也正是由于此原因，两种系统在很多方面是相似的。

Windows 系统可以满足日常的桌面应用，但是无法处理网络重负载。很少有机构尝试让它作为 Internet 服务器来运行，例如 barnesandnoble.com 网站使用 Windows-NT 系统，但是 WEB 服务器经常产生如下的错误信息，"Error Message: [Microsoft][ODBC SQL Server Driver][SQL Server]Can't allocate space for object 'queryHistory' in database 'web' because the 'default' segment is full"。（错误消息：[Microsoft][ODBC SQL Server Driver][SQL Server]不能在数据库'web'中为对象'queryHistory'分配空间，因为'default'字段已满。）微软公司自己的 Hotmail 服务器使用 FreeBSD 系统已经很多年了。

安全性

FreeBSD 系统的安全性已经作为一个课题被大量的研究，所有的可能会导致问题的系统成分都被再三的检验以发现与安全相关的错误。由于整个系统是开放的，所以安全性可以被第三方所检验。FreeBSD 的默认安装毫无疑问受 CERT 2000 年的安全报告影响。

FreeBSD 系统内核分为不同的安全等级，这比单级运行的内核更强大，因为它允许超级用户完全拒绝一些系统操作，像读磁盘/内存，改变文件系统标志，在没有挂接文件系统时对磁盘进行写操作。FreeBSD 系统包含了非常强劲的包过滤防火墙系统和许多非法入侵检测工具。

Linux 系统的开放源代码特性允许任何人对代码的安全性进行检查和修改，但实际上，系统的核心代码被一些经验不足的程序员修改得太快了。由于没有正式的代码检查策略，Linux 系统的安全性被几乎所有的基于

UNIX 系统的 CERT 报告提及。

但是 Linux 系统同样也有很强劲的包过滤防火墙系统和许多非法入侵检测工具。

微软公司声称自己的产品是安全的，但是他们无法提供保证，并且他们的软件在进行测试时经常是不可用的。由于 Windows 系统不开放源代码，因此对于用户来说无法对定期发布的微软系统的安全补丁进行诊断和修改。

设备驱动

FreeBSD 系统允许在自举的时候装载二进制驱动，它允许任何第三方的驱动生产商发布可以在 FreeBSD 系统上运行的驱动程序。由于 FreeBSD 系统是开放源代码的，因此开发新硬件的驱动程序是非常容易的，但是大多数的硬件厂商只提供对微软的操作系统的驱动，因此往往在新硬件上市数月之后才有对 FreeBSD 系统的驱动程序。

Linux 团体刻意给那些提供二进制驱动版本的硬件制造商制造麻烦，也就是他们鼓励硬件制造商提供开放源代码的驱动，但是大多数的硬件厂商不愿意提供开放他们的驱动的源代码，因此对大多数 Linux 使用者来说根本无法使用这些驱动。

微软公司和硬件厂商保持着很好的关系，虽然不同的微软 Windows 版本之间的硬件驱动会存在一些冲突，但是所有的 Windows 用户都可以很好的使用第三方设备驱动。

商业应用

针对 FreeBSD 系统的商业应用程序增长很快，但是相比 Windows 而言还是要差很多。除了自身的应用程序以外，FreeBSD 可以运行针对 Linux，SCO UNIX，and BSD/OS 等系统编译的程序。

许多新的 Linux 系统商业程序已经可以使用，而更多的还正在开发中。但是，由于 Linux 系统只能运行针对 Linux 系统编译的程序，而为 FreeBSD，SCO UNIX 和其他系统编译的程序就无法运行于 Linux 之上了。

Windows 系统有成千上万的应用程序，比其他所有的操作系统都要多，几乎所有的商业应用都运行在 Windows 平台之上，而且大多数都只能运行在 Windows 平台上。如果你的重要的应用软件只能运行在 Windows 平台，你就不得不使用 Windows 系统了。

开发环境

FreeBSD 系统包括完整的开发环境，你可以得到完整的 C/C++开发环境（编辑器、编译器、调试器、模拟器等）和强大的针对 Java，HTTP，Perl，Python，Tcl/Tk，Awk，Sed 等的 UNIX 开发工具，所有这些都是免费的，而且在最基本的 FreeBSD 安装包中都有，并提供了源代码。

Linux 包含了和 FreeBSD 相同的开发工具，为所有的通用程序设计语言提供编译器和解释器，包括 GNU C/C++编译器，Emacs 编辑器，和 GDB 调试器。但是，由于 Linux 的原始版本非常多，因此在某个系统下（如 Red Hat 7）编译的程序可能在其他系统下（如 Slackware）无法运行。

Windows 2000 提供很少的开发工具，大部分需要另外购买，而且很少相互兼容。

支持

FreeBSD 可以在 Internet 上自由下载，或者是购买 4 张 CD 盘，其中包含了几个 GB 的应用程序，售价 40 美元，所有必须的文档都包含在内。技术支持是免费的或者是非常便宜，而且不存在用户许可证，因此你可以在多台计算机上安装并迅速投入应用，这些对用户来说代价非常小。

Linux 也是自由软件，一些公司以很低的价格提供了商业版本，应用程序和文档也是免费的或者非常廉价，没有用户限制，因此 Linux 可以被你随心所欲的安装在多台计算机上。

Windows 2000 的服务器版本的价格将近 700 美元，而一些基本应用程序都需要另外收费。用户通常要花费几千美元购买一些程序软件，这些程序软件在 Linux 或者 FreeBSD 系统中是不收费的。文档非常昂贵，而系统只提供了少量的在线文档。每台计算机都需要授权证书，也就是说会导致延误和增加额外的管理。对于最初掌握简单管理任务的过程，使用 Windows 2000 要比 UNIX 花费的时间少，但是需要增加大量的工作用于保持系统稳定运行。

第 6 章　软件工程

6.1　软件基本概念

软件是一个用于操作计算机及其相关设备的各种程序的总称。名词"硬件"描述了计算机和相关设备的物理概念。

软件可以看作是一个计算机中可变的部分，而硬件是不可变的部分。软件通常可以划分为应用软件（用户直接感兴趣的程序）和系统软件（包括操作系统和许多其他程序）。中间件有时用来指处于应用程序和系统程序之间，或两种不同的应用软件之间的编程（例如，实现将数据从一种文件格式向另一种文件格式的转换）。

另外一种难以分类的软件是实用软件，它是容量有限、但却有用的小程序。一些实用程序捆绑在操作系统中。跟应用程序一样，实用程序一般另外安装，并允许独立于操作系统使用。

按购买或获取的途径，软件可以分为共享软件（通常在试验期后打算销售），不具有部分性能的共享软件，免费软件（免费但拷贝权受到限制），公共软件（不限制拷贝），和用户同意不限制进一步分发销售的免费软件。

软件通常存放在 CD-ROM 或软盘上。现在，很多购买的软件，共享软件和免费软件都可以从网上下载。

6.1.1　应用软件

应用软件可以看作是"终端用户"软件，它是基于通用任务执行的一些有用的工作，像文字和表格处理。

常用的应用软件有：

- 创造性软件，包括字处理软件，电子表格软件，和大多数计算机用户使用的工具。
- 演示软件，像微软的 PowerPoint。
- 绘图设计用的绘图软件，包括 Photoshop，3ds max 等。
- 计算机辅助设计/计算机辅助制造（CAD/CAM）软件。
- 专门的科学应用软件。

应用软件可以是打包的程序或基于用户需要编写的程序。

- 软件包是指那种由专业编程人员预先写好的，制作在软盘上以供销售的程序，仅微型机就有 12000 多种不同种类的应用软件包。
- 曾经所有的软件都是用户定制软件，或称用户软件。二十年前，各公司雇用编程人员编写所有的软件。编程人员根据用户的需要编写程序，指示公司的计算机执行组织想要实现的所有任务。一个程序可以计算工资单，记录仓库里的货物，估算销售任务，或者完成类似的商业任务。

有一些通用程序我们称之为"基本工具"，它们被广泛地应用于几乎所有的行业领域中，要了解计算机就应该知道这些程序。

最流行的基本工具包括：

- 字处理程序：准备书面文档；电子表格：分析和汇总数据；
- 数据库管理程序：组织和管理数据与信息；
- 绘图程序：可视化分析和描述数据与信息；

- 通讯程序：传输和接收数据与信息；
- 集成程序：将一些或所有应用程序集成在一个程序中。

6.1.2 系统软件

用户和应用软件之间相互影响。系统软件使应用软件可以与计算机之间相互影响。系统软件可以看作是一种"背景"软件，包括那些帮助计算机管理自身内部资源的程序。

最重要的系统软件就是操作系统，它工作于应用软件和计算机之间。操作系统处理的是运行（执行）程序，存储数据和程序，处理（操作）数据这样的细节。操作系统将用户从操作计算机的复杂性中解脱出来，使之专注于解决问题。

微型机的操作系统伴随着机器本身的发展也在变化，比早先的操作系统要强大的多。现在操作计算机要求具备下面这些最流行的微型机操作系统知识：

- DOS 是 IBM 公司制造的 PC 机和兼容机上所使用的标准操作系统。
- Windows 最初并不能称为是一个操作系统，而只是由 DOS 扩展功能后的运行环境。
- Windows NT 是一种为功能强大的微型机而设计的操作系统。
- Os/2 是由 IBM 公司为性能更强大的微型机开发的一种操作系统。
- Macintosh 操作系统是苹果公司为其 Macintosh 型计算机设计的标准操作系统。
- UNIX 是一种起初为小型机开发的操作系统。由于它可以运行于许多功能强大的微型机之上，UNIX 是如今重要的操作系统之一。

6.2 软件生存期

早期对软件工程的定义是由 Fritz Bauer 在第一次针对这一学科的重要会议上提出的：

建立和使用稳妥的工程原理，以期得到一种能在实际机器上进行可靠而高效工作的、经济的软件。

虽然已经提出了许多非常全面的定义，但所有的定义都强调软件开发工程规则的重要性。软件工程是硬件和系统工程的派生物，它包含三个关键元素：方法、工具和过程，这些元素使软件管理者能控制软件开发的过程，并为软件具体开发者提供一个建立以多产方式生产高效软件的基础。

图 6-1 给出软件工程中传统的软件生存期模型，有时，称之为"瀑布模型"，它要求软件的开发从系统级设计到分析、设计、编码、测试和维护，都遵循系统的、顺序的方法。仿照传统的工程周期，软件生存期模型包括下列活动：

6.2.1 系统工程与分析

由于软件总是较大系统的一部分，所以软件设计的工作要从为所有系统元素建立需求开始，然后再将这些需求的某一子集分配给软件。当软件必需与其他元素接口时，像硬件、人和数据库，系统观点是不可少的。系统工程和分析系统级的需求收集和少量的高层设计和分析。

6.2.2 软件需求分析

需求分析的收集过程特别重要，并且要着重从软件的角度进行收集。为了理解要创建程序的性质，软件工程师（"分析员"）必须理解软件的信息范畴，以及要求的功能、性能与接口。对系统和软件的需求都要建档，以供用户查阅。

6.2.3 设计

软件设计实际上是一个多步骤的过程，关键在于三个显著特征：数据结构、软件构架，和过程细节。设计的过程是将需求转化为一种在编码开始前可以评估软件质量的表示形式。像需求分析一样，设计也要建档，成为软件配置的一部分。

6.2.4 编码

设计必须翻译成机器可识别的形式,编码阶段的任务就是做这样的工作。如果设计做得很详细,编码可以由机器实现。

6.2.5 测试

一旦代码生成,程序的测试就开始了。测试的过程着重于软件的内部逻辑(确保所有的语句被测试到)以及外部功能(确保所定义的输入能产生出与要求相符的结果)。

6.2.6 维护

毫无疑问,软件在递交给用户之后还要进行修改(嵌入式软件例外)。软件一遇到错误就要改变,或是因为软件必须适应它的外在环境的变化(例如,如果换了新的操作系统或外设产生了变化),或是因为用户需求提高功能和性能。软件维护是将前述的各种生命周期步骤应用于现存的程序。

传统的生存期是软件工程中使用最早的,也是应用最广泛的模型,尽管如此,最近几年出现的对模型法的批评,甚至使那些积极的支持者也质疑它是否适合于所有情况中。下面是传统的生存期模型有时会遇到的问题:

- 真正的项目很少依照模型所提出的顺序流程来实现,时常会出现迭代,这样会在模型的应用中制问题。
- 要求用户在开始时明确指出全部要求往往是困难的。但传统的生存期却要求如此,这样很难适应许多项目在设计之初就存在的不确定性。
- 这种模型要求用户必需有耐心,因为程序的工作版本直到整个项目的后期阶段才可用。到验收程序时才发现的一个大错会是一个灾难。

所有这些问题都是实际存在的,尽管如此,传统的生存期范例在软件工程中仍占有明显的重要位置,它提出了一个包含分析、设计、编码、测试和维护方法的模板。传统的生命周期仍然是软件工程中应用最广泛的程序模型,尽管它有一些不足,但它仍显著好过杂乱无章地进行软件开发。

6.3 原型法

原型法这样处理设计问题:在开始艰难的编码之前,为系统分析员和用户提供一个机会来捕捉设计中的毛病。

当系统分析员和用户设计出一个计划时,首先构造出一个原型。接着,系统分析员利用一个开发工具设计出应用。最后,用户和分析员一起坐在屏幕前,移动、修改元素或突出显示它们,使应用程序获得更好的工作性能。

通常使用通用工具来构建原型,如应用程序发生器。而被称为高效编程工具的其他软件包也适合于这项任务,包括第四代语言、文件发生器和系统开发工具。

由于取消了传统开发系统中常用的纠错式编码,原型法可以直接减少制作应用程序的费用。此外,它可以大大地提高数据中心职员的生产率,间接地节约资金。

一些支持者认为,即使失败的原型也会给用户一个受教育的经历。在创建的过程中,用户得到 MIS/DP 试图解决的第一手问题。许多原型最终被扔进了垃圾箱,但试验是游戏的一部分,不能认为是缺点。公司必须要牺牲一定的时间和付出努力以从原型法中获利,而那些好处要远大于最初失败付出的代价。

当开发过程中采用原型法时,不要试图在几个部门之间处理的系统采用原型。从简单和自制的不必集成到其他系统的事开始,如市场信息系统。

小心地列出项目中用户成员名单,用户成员的级别越高,系统就会做得越好。并不是所有的编程人员和分析员都适合开发原型法，因为他们必须与客户紧密和作，这是一个某些技术人员很难做到的角色。

与用户建立完好关系不会直接解释为根本改善了关系。尽量通过减少生成必不可少的新信息系统所需时间，原型可以帮助公司改善竞争能力。

第 7 章 程序设计语言

7.1 程序设计语言简介

程序设计语言是指令计算机实现某些具体任务的一套特殊词汇和一组语法规则。从广义的角度说，它包括一组既能被人所理解又能被计算机所识别的声明和表达式。为人们所能理解是因为它们使用的是人类的（英文和数学的）表达方式。另一方面，计算机通过使用专门的程序来处理这些指令，这些专门的程序就是我们所熟知的翻译程序，它能解码我们发出的指令并生成机器语言代码。

所谓程序设计语言通常是指高级语言，像 BASIC，C，C++，COBOL，FORTRAN，Ada 和 Pascal。每种语言都具有一套独特的关键字（它能理解的字）和组织程序指令的专门语法。

简单地与人类语言相比，高级程序设计语言比计算机实际识别的语言，也就是机器语言，复杂得多。不同型号的 CPU（中央处理单元）都有它独自的一套机器语言。

处于机器语言和高级语言之间的是汇编语言，它直接与机器语言相关；也就是说，它可将一个汇编指令生成一个机器语言指令。人们几乎不可能去读和写那些只包含数字的机器语言。虽然汇编语言具有和机器语言相同的结构和命令集，但是编程人员可以使用助记符来代替数字。

每种类型的 CPU 都有它自身的机器语言和汇编语言，因此为一种 CPU 编写的汇编语言不能运行于其他的 CPU 之上。在程序设计的早期，所有的程序都是用汇编语言编写的。而现在，大部分程序都是用像 FORTRAN 或 C 这样的高级程序编写的。但当对运行速度要求很高，或执行一个高级语言无法处理的操作时，编程人员仍旧会选择汇编语言。

位于高级语言之上的是第四代程序设计语言（简称 4GL），它与机器语言差异更大，代表了与人类语言更为接近的那类计算机程序设计语言。大多数 4GL 被用于进行访问数据库的操作。例如，一条典型的 4GL 是：FIND ALL RECORDS WHERE NAME IS "SMITH"（查找所有记录中姓名是"SMITH"的记录）。

近来出现的新的程序设计语言，像 C++，微软的 Visual C++，Visual Foxpro，和 Visual Java，他们都支持面向对象的程序设计（OOP）技术。OOP 这种程序设计要求编程人员不仅要给出数据结构中的数据类型的定义，还需要给出作用在这些数据结构之上的操作（函数，或方法）的类型。这样，可使得数据结构成为一个既包含数据又包含函数的对象。此外，编程人员可以创建对象间的关系，例如，一个对象可以继承其他对象的一些特性。相对于过程式的程序设计技术，面向对象的程序设计技术的一个主要优点是：当创建一个新类型的对象时，已有的模块不需要发生改变。编程人员可以很方便地创建一个新对象，使它的许多特征继承于已存在对象，这使得面向对象的程序更容易修改。为了运行面向对象的程序设计，需要一种面向对象的程序设计语言（OOPL）。C++和 Smalltalk 是其中最流行的两种语言，当然也有 Pascal 的面向对象的版本。

所有的高级语言程序只有在翻译成机器语言后，才能被计算机执行。翻译有两种方式：编译和解释。

到底哪种语言最好是一个困扰业界专家很长时间和消耗他们很多精力的问题。每种语言都各有优劣，例如，像 FORTRAN 语言，适合于处理数值计算，但却不适合组织大规模的程序。Pascal 语言是一种结构化的语言，程序可读性强，但却不具备 C 语言的灵活性。C++嵌入了功能强大的面向对象特性，但是它却复杂难学。具体选择使用哪一种语言依赖于程序运行的计算机类型，程序的种类，和编程人员的专长。

设计更高级的语言的趋势始于 20 世纪 50 年代并一直进行到了现在。如今，人们在探讨人与计算机之间进行自然语言的交流的问题；也就是说，正在研制的工具使得人们可以用正常的、口语式的语言同计算机讲话，不论是书面的形式还是口头的形式。然而，要想这些语言能在商业中广泛应用，还需要做很多努力。

7.2 面向对象的程序设计

面向对象的程序设计是一种以面向"对象"的，而不是"行为"的，面向数据而不是逻辑的方式组织的程序设计语言模式。曾经，程序被认为是接收输入数据、处理、产生输出数据的一个逻辑过程。编程的困难在于如何写出逻辑关系，而不是如何定义数据。而面向对象的程序设计认为，我们真正关心的是我们所能操作的对象而不是操作它们所要求的逻辑关系，比如说，从人（由姓名、地址，等等来描述）到建筑物和楼层（可由它的特征来描述和管理），再到你的计算机桌面上的小控件（像按钮和滚动块）。

面向对象的程序设计的第一步就是要确定你需要操作的所有实体和它们之间的联系，一个为人熟知的操作就是数据建模。一旦你定义了一个对象，你将它概括为一个对象的类（试想，柏拉图给出的"完美"椅子的概念代表了所有的椅子）并定义了这个类所包含的数据和所有可操作的逻辑顺序。每一个明确的逻辑过程就叫做方法。一个类的示例（这里毫不奇怪地）叫做一个"对象"，有时也叫做"一个类的示例"。机器上运行的就是对象或类的示例。其中，方法给计算机提供指令，类和对象的特征提供相关的数据。你可以和对象交流，而对象之间通过定义好的接口，也称为消息，进行交流。

面向对象的程序设计中的概念和规则具有下面这样一些重要的优点：

- 数据类的概念使得定义的子类可以部分地或全部地共享父类的特征。这称之为继承。OOP 的这种特性要求必须深入的分析数据，减少研发的时间，确保更准确的编码。
- 由于一个类只需定义与之相关的数据，当这个类的一个实例（对象）运行时，它的代码不能突然去访问其他程序的数据。这种数据的隐藏性提供给系统更强的安全性，也避免了非故意的数据破坏。
- 一个类的定义不仅可为最初创建它的程序再使用，也可以被其他的面向对象程序使用（也正是这个原因，类适用于分布式网络）。
- 编程人员利用数据类的概念可以创建任何计算机语言中没有定义过的数据类型。

Smalltalk 是最早的面向对象的计算机语言之一，而如今最流行的面向对象语言是 **C++** 和 **Java**。**Java** 语言最适用于私网和因特网上的分布式应用。

7.3 OMG 的统一建模语言

大型企业应用软件——那些维持公司正常运行的核心业务应用程序，绝不仅仅是一大堆的软件模块。它们必须是在紧迫环境下具备可扩展、安全性、强壮性的结构。其构造——通常称为体系结构，必须明确地定义，使得原始编码员转去开发其他项目以后，维护程序员也能快速发现并修补出现的错误。这就是说，这些程序的设计必须很好地适应不同领域，且业务功能不是单一的（不用说，它当然是基本核心）。当然，一个好的体系结构对任何程序都是有益的，并不限于我们这里提到的大型应用软件。我们首先提到大型应用，因为结构化是处理复杂性的方法，所以结构化的益处随着应用规模的增大而增大。结构化的另一益处是代码重用。将一个应用按内置的模块或组件构造的话，设计就是很容易的事了。最后，企业建立了一个组件模型库，每一个组件代表存在库中的代码模块的实现。当另一应用需要相同功能时，设计者能够快速地从库中导入相应模块。在编码阶段，开发者就能快速导入代码模块执行。

建模是应用软件在编码前的设计是大型软件项目的基本部分，也同样适于中型甚至小型项目。一个模型在软件开发中的作用类似于在建筑一座摩天大楼中蓝图和其他计划（现场地图、建筑物的正视图、物理模型）

所起的作用。使用模型，那些负责软件开发项目的人能确保业务功能完备、正确，能满足端用户的需求，程序设计支持对可伸缩性、鲁棒性、安全性、可扩展性以及其他特性的要求，在代码实现前开发者就改变了以往代码修改的困难和昂贵代价。调查表明，大型软件项目的失败率很大。事实上，在给定时间和预算内，大型软件项目很可能因满足不了全部需求而失败。假设需要开发类似这样的大项目，必须竭尽所能提高成功的几率，而模块化是唯一可选。模块化可令员工在开始编码前就了解原创者的设计并且检验需求是否满足。

OMG 的统一建模语言（UML）可在满足所有需求的前提下，确定并直观地描述软件系统的模型，包括系统的结构和设计（可采用 UML 进行商业建模，也可将之用于其他非软件系统的建模）。用户可选用市场上提供的众多 UML 工具的任一种，分析自己未来应用的需求，并且设计一个方案来满足这些需求，结果是用 UML 的十二类标准图表来描述的。

UML 可为任意类型的应用建模，而不管它是运行在哪一类硬件或硬件组合上，使用哪一种操作系统、编程语言和网络。UML 的灵活性确保它能为使用任一种中间件的分布式应用建模。由于是建立在 MOF 原型基础上，以类和操作作为基本概念，UML 非常适于面向对象的语言环境，如 C++、JAVA 和最新出现的 C#。然而，UML 同样可为非面向对象的应用建模，如 Fortran、VB 或 COBOL。UML 协议子集（即：用于特定应用的 UML 子集）有助于以一种更自然的方式为事务性实时容错系统建模。

UML 还能做其他有益的事。例如：UML 的某些工具分析现有的源代码并把它反编辑成一组 UML 图表。还有其他的例子：尽管 UML 侧重设计而非执行，也有些可用工具能执行 UML 模型，代表性的两个方面是：某些工具通过一种方法解释执行 UML 模型，使你确信它做到了你所需要的，但还不具备开发软件中应有的可扩充性及速度要求；另一些工具（通常仅限于有限的应用范畴，如电信、金融领域）从 UML 生成编程语言代码。如果代码生成器并入训练好的可扩充模式，如：事务性数据库操作或其他一般的程序任务，那么产生的大多是无错的、可快速运行的应用。还需说明的是：市场上的大量工具从 UML 模型中生成测试和验证软件包。

几年前，开发者在开发一个分布式程序的项目时遇到的最大问题是要找到一个具备所需功能的中间件，可以在他店里的硬件和操作系统上运行。而今天，面对着一个丰富到令人眼花缭乱的中间件平台市场，开发人员有三个不同的难题：第一，选择哪一种；第二，中间件要能在不同的平台上工作，不仅限于自己的店，还要在其客户和供应商的那些已经配置好的平台上工作；第三，当一个新的平台出现，并且符合分析家的设想，要有和新的"下一个最好的工具"的接口，CIO（美国产业联合会）标准是必不可少的。

第 8 章　因特网

因特网是什么？它来自何处又是如何支持万维网的增长？因特网最重要的运行原则是什么？因特网是成千上万的网络和数以百万计的计算机（有时被称为主机计算机或主机）将企业、教育机构、政府机关和个人联结起来的一个互联网络。因特网为全球大约 4 亿人（其中美国 1.7 亿多人）提供诸如电子邮件、新闻讨论组、购物、研究、即时信息、音乐、视频和新闻等服务。尽管它为商业交易、科学研究和文化提供基础设施。没有任何组织控制或运行因特网，它也不被任何人所拥有，因特网（Internet）一词起源于互联网络（internetwork）或两个或更多的计算机网络联接在一起。全球信息网，或简称万维网，是因特网上最流行的服务之一，提供对 10 亿多网页的访问，这些网页是由一种叫做超文本链接标示语言（HTML）生成的文件，它可以包含本文、图形、声频、视频和其他对象、以及允许用户容易地跳跃到其他网页的"超链接"。

8.1　因特网：主要技术概念

在 1995 年，联邦的网络委员会（FNC）通过了一个关于因特网术语正式定义的决议。

"因特网"是这样的全球信息系统：
- 通过以因特网协议或其扩展或继续为基础的、独特的地址空间被逻辑性地联结起来；
- 能够支持使用 TCP/IP（传输控制协议/因特网协议）组或其扩展/继续、和/或其他与 IP 协议（因特网协议）兼容的协议进行的通信；
- 提供、使用或使可访问此处描述的通信及其相关基础设施（不论公用的还是专用的）上分层次的高水平服务。

基于这个定义，因特网表示这样一个网络——使用 IP 地址分配方案、支持传输控制协议，并使用户可以使用多种服务，这与电话系统使公众能够使用声音和数据服务非常类似。

在这个正式的定义背后，隐含着三个极其重要的概念：分组交换、TCP/IP（传输控制协议/网际协议）通信协议和客户机/服务器计算技术，它们乃是理解因特网的基础。尽管因特网在过去 30 年发生了引人注目的进化和变化，但这三个概念仍是今天因特网运转的核心，也是将来因特网的基础。

分组交换。分组交换是传输数据的一种方法，它先将数据信息分割成许多称为"分组"的数据信息包；当路径可用时，经过不同的通信路径发送；当到达目的地后，再将它们组装起来。在分组交换发展之前，早期计算机网络使用租用的专用电话线路和终端与其他计算机进行通信。在线路交换网络（如电话系统）中，一个完全点对点的线路被连结在一起，然后才能进行通信。然而，这些"专用的"线路交换技术既价格昂贵还浪费有效的通信能力——不论是否有数据输送都需要维持线路。由于字间的停顿和组装时的延迟，一条专用的声音线路在几乎 70%的时间内没有得到充分利用，而这两种因素都增加寻找和连接线路所需的时间长度。因此需要一种比较好的技术。

第一本关于分组交换的著作是由 Leonard Kleinrock 于 1964 年所著，美国和英国防卫研究实验室的其他研究人员使这项技术得到进一步发展。由于使用分组交换技术，网络的通信能力提高了 100 倍甚至更多。数字网络的通信能力用位/秒来衡量。想象一下汽车行驶的里程，从每加仑汽油行驶 15 里提高到每加仑汽油行驶 1500 里——而汽车没有太大的改变！

在分组交换的网络中，信息首先被分解为许多信息包。每个信息包附加数字代码用于指示其源地址（开始点）和目的地地址、以及顺序信息和错误控制信息。在分组网络中，信息包不是直接被送到目的地地址，而是在计算机与计算机之间旅行直到它们到达目的地。这些计算机叫做路由器。路由器是一种特殊用途的计算机，它将组成因特网的成千上万个不同计算机网络互相联接起来，并在信息包旅行时将它们向终极目的地发送。路由器使用一种叫做路由算法的计算机程序，以确保信息包取通向它们目的地的最佳可用路径。

分组交换不需要一个专用线路，但是可以利用几百条线路中任何可用的空闲能力。分组交换几乎充分利用了所有可用的通信线路和能力。而且，如果一些线路不通或太忙的话，信息包能在任何可用的、最终通向目的地的线路上传送。

TCP/IP。尽管分组交换是通信能力的一个巨大进步，但对于将数字信息分解为信息包、将它们传输到适当的地址，然后重新组装为原来的信息，还没有一种公认的方法。这就像有了一个生产邮票的系统，而没有邮政系统（一系列的邮局和一套住址）一样。

TCP/IP 回答了在因特网上用信息包做什么和如何处理信息包的问题。TCP 指传输控制协议，IP 表示网际协议。协议是一组用于信息的格式化、次序化、压缩和检查错误的规则。它也可以限定传输速度和网上设备显示它们已停止发送及（或）接收信息的方法。协议既可以通过硬件也可通过软件来实现。TCP/IP（传输控制协议/网际协议）通过被称为服务器软件的网络软件来实现（在下文描述）。TCP（传输控制协议）是用来在网上传输数据的公认协议。TCP（传输控制协议）在发送和接收网络计算机之间建立连接，处理信息包在传输点的组装和在接收端的重新组装。

TCP/IP（传输控制协议/网际协议）被分为四个独立的层，由每层处理一个通信问题的不同方面。网络接口层负责信息包在网络媒体上的排列和接收，网络媒体可能是局域网（以太网）、令牌环形网或其他网络技术。TCP/IP（传输控制协议/网际协议）独立于任何局部网络技术，并能适应在局部水平上的改变。因特网层负责信息的寻址、封装及其在因特网上的路线排定。通过对信息包去与来应用层的确认和排序，传送层负责为应用层提供通信。应用层为许多应用提供访问较低层服务的能力。一些众所周知的应用是超文本传输协议（HTTP）、文件传输协议（FTP）和简单邮件传输协议（SMTP），稍后我们将在本章讨论。

IP 地址。TCP（传输控制协议）处理因特网信息的分组化和传输路线排定。IP（网际协议）提供因特网的地址分配方案。每部连接到因特网的计算机必须分配一个地址，否则它不能够发送或接受 TCP（传输控制协议）信息包。举例来说，当你在使用调制解调器拨号上网时，你的计算机由因特网服务提供商分配一个临时地址。

因特网地址，即众所周知的 IP 地址，是一个 32 位的数字，它以一串由园点隔开的四个数字出现，例如 201.61.186.227。四个数字中的每一个都在 0~255 范围内。这个"点分四元组"地址分配方案包含多达到 40 亿个地址（2^{32}）。最左边的那个数字指示计算机的网络地址，而其余的数字帮助识别正在发送（或接收）信息的团体里面一台特定的计算机。

IP 现在版本叫做第 4 版，或 IPv4。因为许多大公司和政府在各自的域中已给定了数百万个 IP 地址（以适应他们当前和未来的劳动力），且由于新网络和新的可上网设备需要独特的 IP 地址，新版本 IP 协议 IPv6 正在得到采用。这个方案包含 128 位的地址，或曰大约 10^{15} 个地址。

图 8-1 说明了在因特网上发送数据时，TCP/IP 和分组交换是如何协同工作的。

域名和统一资源定位（URL）。大多数人记不住 32 位的数字。IP 地址可由一个自然语言约定（称为域名）来表示。域名系统（DNS）允许像 cnet.com 这样的表达代表数值型 IP 地址（cnet.com 的数值型 IP 地址是 216.200.247.134）。统一的资源定位是网络浏览器用于识别网上内容位置的地址，也使用域名作为它的一部分。一个典型的网址包含访问地址时使用的协议，接着是它的位置。例如，网址 http://www.azimuth-interactive.com/flash_test 就是指 IP 地址 208.148.84.1，其域名为"azimuth-interactive.com"，而访问地址时使用的协议为超文本传输协议（HTTP）。一种叫做"flash_test"的资源位于服务器目录路径/flash_test 之上。一个统一资源定位可以有二到四个部份，例如：name1.name2.name3.org。我们将在下一章中进一步讨论域名和统一资源定位。表 8-1 概括了因特网地址分配方案的重要成份。

表 8-1　因特网的难题：域名与地址

IP 地址	每部连接到因特网的计算机必须有一个称为 IP 地址的一个独特的地址数字。甚至一台使用调制解调器的计算机也被分配暂时 IP 地址
域名	域名系统允许如 aw.com（Addison Wesley 的网址）这样的表达式代表数值型 IP 地址
域名系统服务器	域名系统服务器是记录因特网上 IP 地址和域名的数据库
根域名服务器	根域名服务器是一个中央目录，它列出正在使用的所有域名。当排定数据传输路线时，域名系统服务器从根服务器查阅不熟悉的域名
因特网地址与域名管理委员会	因特网地址与域名管理委员会建于 1998 年，旨在建立域名与 IP 地址规则，并协调根服务器的运行。它从诸如 NetSolutions.com 等私人公司接任

客户机/服务器计算技术。虽然分组交换使有效的通信能力激增，且 TCP/IP 协议提供了通信规则，在计算机技术中，发生了又一次革命才造就了今天的因特网和万维网。这次革命叫做客户机/服务器计算技术，没有它，万维网及其丰富的信息将不会存在。事实上，因特网是客户机/服务器计算技术的一个巨大实例，其中，

包括超过 7000 万部主机服务器、计算机储存网页和其他内容，这些网页和内容能被全世界接近一百万个局域网和数亿台客户机容易地访问。

客户机/服务器计算技术是一种计算模型，其中相当多台被称为客户机的个人计算机联结在一起并与一台或更多的服务器计算机联结在一个网络中。这些客户机功能足够强大以完成复杂的任务，如显示丰富的图形、储存大型文件、处理图形和声音文件，这些任务全部在当地的台式或手持式装置上完成。服务器是联网的计算机，专门用于提供客户机在网上需要的公共功能，例如储存文件、软件应用、公用程序如网络连接、和打印（见图 8-2）。

为了充分理解客户机/服务器计算技术的作用，你必须了解在它之前的东西。在 20 世纪 60 年代和 70 年代的大型计算机环境中，计算机能力非常昂贵而有限。比如，60 年后期最大的商业化大型机有 128K 随机存取存储器和 10 兆磁盘驱动器，而且占据数百平方英尺空间。计算能力不足以支持图形和文本中的颜色，更不用说声音文件或超链接文档和数据库。

20 世纪 70 年代后期和 80 年代初期，随着个人计算机和局域网的发展，客户机/服务器计算技术成为可能。客户机/服务器计算技术比中央大型机计算有许多优势。比如，通过增加服务器和客户机，可很容易地扩容。同时，客户机/服务器网络比中央计算结构不易受损。如果一个服务器被破坏，备用的或镜像服务器能恢复工作；如果一部客户机不能运转，网络的其部分可继续运行。而且，负荷处理通过许多强大的、较小的机器达到平衡，而不是集中在一部为每个人进行处理的巨型机。在客户机/服务器环境中，建立软件与硬件都更简单、更经济。

8.2 其他因特网协议及应用程序

有许多其他因特网协议以因特网应用的形式为用户提供服务，这些因特网应用在客户机和服务器上运行。这些因特网服务以全世界普遍承认的协议或标准为基础，对任何使用因特网的用户有效。它们不为任何一个组织所拥有，而是一种发展了许多年、供给所有因特网用户的服务。

HTTP：超文本文档。HTTP（超文本传输转移协议的缩写）是用于传输网页的因特网协议（下文描述）。HTTP 协议在 TCP/IP 模型中的应用层上运行。当一个客户机的浏览器向远程因特网服务器的一个网页发出请求时，HTTP 会话开始。当服务器通过发送所请求的页面回应时，关于这个对象的 HTTP 会话结束。因为网页可能带有许多对象——图形、声音或视频文件、帧等——每个对象必须由一个独立的 HTTP 信息所请求。

SMTP、POP 和 IMAP：发送电子邮件。电子邮件是最老、最重要和经常使用的因特网服务之一。STMP（简单邮件传输协议）是用于将邮件发送到服务器的因特网协议。客户机利用 POP（邮局协议）从因特网服务器上检索邮件。你可以通过察看浏览器上"收藏"和"工具"栏看见你的浏览器如何处理 SMTP 和 POP 协议，邮件设定在这两栏定义。你可以设定用 POP 协议取回来自服务器的电子邮件信息然后删除服务器上的信息，或在服务器上保留它们。IMAP（因特网消息访问协议）是一个由许多服务器和所有的浏览器支持的更通用的电子邮件协议。IMAP 允许使用者在从服务器下载邮件之前搜寻、组织和过滤邮件。

FTP：传输文件。FTP（文件传输协议）是最初的因特网服务之一。它是 TCP/IP 协议的一部份，它允许用户把文件从服务器传送到客户机，反之亦然。文件可以是文档、程序、或大型数据库文件。FTP 是传输大于 1M 文件最快速和最方便的方法，而许多邮件服务器将不接受大于 1M 的文件。

SSL:安全性。SSL（加密套接字协议层）是一个在 TCP/IP 中运行于的传输层和应用层之间的、确保客户机与服务器之间通信安全的协议。通过诸如信息加密和数字签名等多种技术，SSL 促进安全的电子商务通信和支付。

Telnet（远程登录）：远程运行。Telnet 是一个在 TCP/IP 中运行的终端模拟程序。用户可以在其客户机运行 Telnet。当用户这么做的时候，用户的客户机模拟一个主机计算机终端（定义于大型机计算时代的工业标准终端是 VT-52、VT-100 和 IBM 3250）。然后将能接入到一台支持 Telnet 因特网的计算机上，从那部计算机运行程序或下载文件。Telnet 是允许用户在远方的一部计算机上工作的第一个"远程工作"程序。

Finger：发现人们。通过使用 Telnet 连接到一个服务器上，然后在提示符的键入"Finger"，你能发现谁登录到远程网络。Finger 是一个 UNIX 计算机支持的应用程序。当由远程计算机支持时，Finger 能告诉你谁登录了、他们已经连上多久和他们的用户名。显然，支持 Finger 会涉及安全问题，所以大多数因特网主机今天不支持 Finger。

Ping：检测地址。你可以"Ping"一台主机计算机来检查你的客户机和服务器之间的连结。给出你有关服务器和因特网速度的一些建议，Ping（Packet InterNet Groper，信息包因特网搜索）程序还将会告诉你服务器回应所需的时间。你可以在一部具有 Windows 操作系统的个人计算机上在 DOS 提示符下键入:Ping<域名>。

Tracert：检查路径。Tracert 是几种追踪路径的公用程序之一，它允许用户跟踪其在因特网上从客户机向远程计算机发出的信息的路径。

8.3　因特网服务提供商（ISP）

在多层次因特网体系中通过向家庭出租因特网访问通道并提供最低水平服务的公司、小型企业，和一些大机构叫做因特网服务提供商（ISP）。因特网服务提供商是零售型供给者——他们为街头、家庭、企业办公室处理"最后一步服务"。约 4500 万个美国的家庭通过全国性的或地方性的 ISP 连接到因特网。因特网服务提供商又利用高速电话或电缆线（高达 45 Mbps）典型地连接到因特网和城域交换站或网络访问节点。

在美国主要的因特网服务提供商，如美国在线、微软 MSN 网络和 AT&T 世界网络和约 5000 个地方性因特网服务提供商，业务范围包括提供拨号上网和数据用户专线电话访问的地方性电话公司，到提供海底电缆调制解调器服务的电报公司，到为一个小镇、城市甚至全国服务的小型"妈妈爸爸"因特网商店提供拨号上网服务。如果你的家庭或小企业需要因特网访问，一个因特网服务提供商将会为你提供服务。

因特网服务提供商提供两种服务:窄带和宽带。窄带服务是目前以 56.6 Kbps 运行的传统电话调制解调器连接（虽然由于线路噪音造成大量的信息包重新发送，真正的吞吐量徘徊在 30Kbps 附近）。这是全世界最普通的连接形式。宽带服务以数字用户专线、电缆调制解调器，电话（T1 和 T3 线）和人造卫星技术为基础。在因特网服务的环境中，宽带指的是任何允许客户机以可接受的速度运行流型声频和视频文件的通信技术，通常可接受的速度指在 100 Kbps 以上。

术语 DSL 指数字用户专线服务,是通过在你家或企业所见到的普通电话线传送高速访问的一种电话技术。服务水平大约从 150Kbps 到高达 1Mbps。数字用户专线服务要求客户居住在临近电话交换中心 2 英里之内（大约 4000 米）。

电缆调制解调器是一种在向家庭提供电视信号的视频电缆上搭载对因特网进行数字式访问的有线电视技术。电缆调制解调器服务范围从 350 Kbps 到 1Mbps。如果临近有许多人同时登录并需要高速服务，缆传服务速度可能降低。

T1 和 T3 是数字通信的国际电话标准。T1 线提供保证 1.54Mbps 的传送，而 T3 线提供 43Mbps 传送。T1 线每月费用约为 1000～2000 美元，而 T3 线费用则在 10000～30000 美元之间。这些租用的、专用的、有保障的线路适用于公司、政府机构和企业（如需要高速度保证服务水平的因特网服务提供商）。

一些人造卫星公司正在为配置了 18″人造卫星天线的家庭和办公室提供因特网内容的宽频高速数字下载服务。可用的服务从 256 Kbps 到 1Mbps。因为他们是单向的，即你能从因特网高速下载，但是不能上传内容到因特网。所以从总体上说，人造卫星连接家庭和小企业是不可行的，对于他们的上传需要，使用者需要的是电话或电缆连接。

第 9 章　万维网

9.1　超文本

因为操作个人计算机的网络浏览器软件能使用 HTTP 协议对储存在因特网主机服务器上的一个网页发出请求，所以能通过因特网访问万维网网页。超文本是使用嵌入式链接形成格式化网页的一种方法，这些链接将文档彼此联结，而且将网页链接到其他对象如声音、视频或动画文件。当你点击一个图形和一个视频剪辑播放按钮的时候，你在点击一个超链接。例如，当你在浏览器中键入一个网址，如 http://www.sec.gov 的时候，你的浏览器对 sec.gov 服务器发出一个 HTTP 请求，请求 sec.gov 的首页。

HTTP 是每个网址的第一个字母组合，位于网址的起始位置，紧跟着它的是域名。域名指定组织的服务器计算机，而文件收藏于服务器计算机之上。大多数公司有一个与其官方公司名字相同或接近的域名。目录路径和文件名在网址中是出现较多的两个信息，它帮助浏览器捕捉被请求的网页。同时，网址叫做统一资源定位符，或 URL。当把网址键入一个浏览器内时，URL 可以准确分辨到哪里找寻数据。举例来说，在下面的网址中：

http://www.megacorp.com/content/feature/082602.html

http=用于显示网页的协议

www.megacorp.com=域名

content/feature=目录路径，识别网页被储存在域名网络服务器上的位置

082602.html=　文件名及其格式（html 页）

当前可以使用的最通用的、并被因特网域名与地址管理委员会正式认可的域扩展名见下面列表。国家也有域名如 .uk、.au 和 .fr（英国、澳洲和法国）。在下面列表中还有最近被核准的顶级域名 .biz 和 .info 和正在考虑的域。在不久的将来，这个列表还将扩大到包括更多类型的组织和行业。

.com　商业的组织/企业

.edu　教育机构

.gov　美国政府机关

.mil　美国军队

.net　网络计算机

.org　非营利性组织和工业标准组织

2001 年 5 月 15 日核准的新顶级域：

.biz　商业公司

.info　信息供给者

计划中的新顶级域：

.aero　航空运输工业。

.coop　合作团体

.museum　博物馆

.name　个人

.pro　专业人士

9.2 置标语言

虽然最常见的网页格式化语言是 HTML，实际上文档格式的概念早在 20 世纪 60 年代随着通用置标语言（GML）的发展就形成了。

SGML（标准通用置标语言）。在 1986 年，国际标准组织正式通过了 GML 的一种变体叫做标准通用置标语言，或 SGML。SGML 的目标是帮助特大型组织对大量文档格式化并归类。SGML 的优点是它能独立于任何软件程序运行，但不幸的是，它极端复杂和难学。或许因为这个理由，它没有被广泛地采用。

HTML（超文本链接标示语言）。HTML 是相对容易使用的一种 GML。HTML 为网页设计者提供一组固定的标示"标签"用于格式化网页。当这些标签被插入一个网页之中时，他们被浏览器阅读并翻译为网页显示。你可以看见任何网页的 HTML 源代码，只需点击一下所有浏览器中都能找到的 "网页源文件"指令。

HTML 的功能是定义文件的结构和风格，包括标题、图形定位、表格和本文格式。从 HTML 引入以来，两个主要的浏览器——Netscape（网景公司）的 Navigator 和 Microsoft（微软公司）的 Internet Explorer——不断地把特征加入 HTML 之中，使程序员能够进一步改进他们的网页设计。不幸的是，许多功能的加强只有在一个公司的浏览器中起作用，而且这种发展威胁到通用计算机平台的实现。更糟糕的是，建立具有专有功能的浏览器使建设电子商务网站的费用增加。每当你建立一个电子商务网站，必须特别注意确保网页能用主要浏览器观看，甚至过时的浏览器版本。

HTML 网页可以使用微软的 Word 或几种网页编辑器中任何一种通过本文编辑器生成，例如 Notepad（记事本）或 Wordpad（写字本）（只需把 Word 文件保存为一个网页）。

XML（可扩展标示语言）。XML 使网络文件格式化发生了一次巨大的飞跃。XML 是由万维网联合会开发的一种新型置标语言规范。XML 是像 HTML 一样的一种置标语言，但是它又有非常不同的目的。XML 是用于描述数据和信息，而 HTML 的目的是控制"表现与感觉"并将数据在网页上显示。

比如，如果你想要把一个病人的病历卡——包括诊断、个人身份、病史信息和任何医生笔记——通过网络从一个在波士顿的数据库发送到位于纽约的一所医院，使用 HTML 是不可能的。然而，使用 XML，这些有关病人的富文件（数据库记录）可以容易通过网络发送并显示。

XML 是"可扩展的"，意味着用于描述并显示数据的标签被使用者定义，而在 HTML 中标签是有限制的，而且是预先定义的。XML 也能把信息转变成新的格式，例如从一个数据库输入信息并显示为一张表格。使用 XML，数据能被有选择性地分析且显示，使它成为比 HTML 更有力的可选方案。这意味着商业公司，或整个行业，全部能使用与网络兼容的置标语言通过发票、可支付帐户、薪资记录和财政信息来描述。一旦完成描述，这些商业文件可以被储存在内联网网络服务器上并在整个公司共享。

XML 仍然不能替换 HTML。当前，只有微软的 Internet Explorer 5 完全支持 XML，而 Netscape（网景公司）不支持（虽然这种现象可能改变）。XML 能否最终替代 HTML 作为标准的网络格式规范，在很大程度上仰赖于将来的网络浏览器是否支持它。目前，XML 和 HTML 在相同的网页上并肩工作。HTML 用来定义应该如何格式化信息，而 XML 用来描述数据它本身。

9.3 网络服务器与客户机

我们已经描述了客户机/服务器计算技术及其在计算机技术结构中引起的革命。你已经知道服务器是一台联接到网络的计算机，用于储存文件、控制外设、与外界——包括因特网接口，且为网络上的其他计算机进行一些处理。

但什么是网络服务器？网络服务器软件是一种使计算机能够向网上客户机传递 HTML 网页的软件，而客户机通过发出 HTTP 请求申请这种服务。网络服务器软件的两种主要品牌是 Apache 和微软的 NT 服务器软件，前者是一种免费的网络服务器共享软件，约占有 60%的市场；后者约占有 20%的市场。

除了回应网页请求之外，所有的网络服务器还提供一些附加的基本能力，例如下列各项：

- 安全服务——主要由认证服务组成，确认人们试图对网站进行的访问是经过授权的。对于处理支付交易的网站，网络服务器也支持加密套接字协议层（SSL）——用于在因特网上安全地传输与接收信息的因特网协议。当私人信息如姓名、电话号码、地址和信用卡数据等需要向一个网站提供时，网络服务器使用 SSL 确保从浏览器到服务器来回传递的数据不被损害。
- 文件传输协议（FTP）——这个协议允许使用者从服务器来回移动文件。根据使用者的身份不同，一些网站限制文件上传到网络服务器，而其他网站限制下载。
- 搜索引擎——正如搜索引擎网站使用户能够为特殊的文件搜寻整个万维网，基本网络服务器软件包中的搜寻引擎模块允许对网站的网页和内容进行索引，并允许网站内容的关键词搜索。当进行搜寻时，搜索引擎使用索引，索引是服务器上所有文件的列表。将搜寻项与索引进行比较，确定可能的匹配。
- 数据捕获——网络服务器还有助于监测网站访问量，捕获有关谁访问某个网站、用户在那里停留多久、每次访问的日期和时间、和服务器上哪个特定网页被存取等信息。这个信息被汇编并保存在一个日志文件中，然后能通过用户日志文件进行分析。通过分析一个日志文件，网站管理员能找出访客总数、平均访问时间长度和最流行的目的地或网页。

术语网络服务器有时也被用于指运行网络服务器软件的实际计算机。网络服务器计算机的领先制造商是 IBM（国际商用机器公司）、Compaq（康柏）、Dell（戴尔）和 Hewlett Packard（惠普）。虽然任何个人计算机都能运行网络服务器软件，但是最好使用一部为这个目的最佳化的计算机。作为一个网络服务器，一部计算机必须安装上述网络服务器软件且联接到因特网。每部网络服务器计算机有一个 IP 地址。例如，如果你在浏览器中键入 www.aw.com/laudon，浏览器软件向域名为 aw.com 的网络服务器发出 HTTP 服务请求。然后服务器找出它硬盘上名为"laudon"的网页，把网页送回到你的浏览器并在屏幕上显示。

除了一般的网络服务器软件包之外，网络上其实还有许多类专用服务器，例如访问数据库特定信息的数据库服务器，传递特定标题广告的广告服务器，提供邮寄信息的邮件服务器，以及提供视频剪辑的视频服务器。在一个小型电子商务网站，所有的这些软件包可能在一部具有一个处理器的机器上运行。在一个大公司的网站，可能有数百台分布的机器，多数具有多个处理器，运行上述特定的网络服务器功能。

另一方面，网络客户机是指任何一种联接到因特网、能够发出 HTTP 请求并显示网页的计算装置。最常见的客户机是采用 Windows 操作系统的个人计算机或麦金托什机，各种不同风格的 UNIX 机远远排在第三。然而，增长速度最快的一类网络客户机并不是计算机，而是个人数字助理（PDA）例如 Palm 和 HP Jornada（掌上电脑）以及配备无线网络访问软件的蜂窝电话。总之，网络客户机可以是任何装置——包括电冰箱、加热器、家用照明系统或汽车仪表盘——只要它能从网络服务器发送并接收信息。

9.4 网络浏览器

网络浏览器的主要目的是显示网页，但浏览器还有更多的功能，例如收发电子邮件和加入新闻讨论组（在线讨论组或论坛）。

目前，94%的网络用户使用 Internet Explorer 或 Netscape Navigator 浏览器，但是最近已开发出的一些新浏览器开始吸引人们的注意力。Oprea 浏览器正在变得非常流行，因为它的速度——它是目前世界上最快的浏览器，而且因为它比现有浏览器小得多（它几乎可以装在一张软盘上）。它也能记得你访问过的最后一个网页，因此下回漫游时，你能从你停止的地方开始。而且像前面那两个浏览器一样，用户能免费下载；其弊端是用户不得不忍受一个角落中可恶的广告，否则要支付 40 美元购买无广告版。

浏览器 NeoPlanet 也正在获得新的用户，主要因为它带有 500 多套界面或设计方案。使用不同界面，你能根据你喜欢看或听的方式设计浏览器，而不是被限于 Netscape Navigator 和 Internet Explorer 浏览器提供的标准外观。然而，NeoPlanet 需要 Internet Explorer 的技术来运行，因此你也必须在计算机上安装 IE。

第 10 章　网络安全

10.1　安全网络和安全策略

怎样才算得上一个安全的网络呢？怎样才能使一个网络变得更安全呢？尽管安全网络的概念对大多数用户都很有吸引力，但是网络并不能简单地划分为安全的或是不安全的，因为安全本身不是绝对的，每个团体对拒绝或允许访问定义了不同的等级。比如，有些单位的数据是很有保密价值的，他们就把网络安全定义为外界不能访问其计算机；有些单位需要向外界提供信息，但禁止外界修改这些信息，他们就把网络安全定义为数据可以被外界任意访问，但不允许未经授权的修改；有些单位注重通信的隐秘性，他们就把网络安全定义为信息不可被他人截获或阅读；有些大的组织对安全的定义会更复杂，他们允许外界访问一些公开的数据和服务，同时有些敏感的数据和服务对外界保密，不允许访问或修改。

正因为安全网络不存在一个绝对的定义，任何组织实现安全系统的第一步就是要制定一个合理的安全策略。该策略不是去限定具体的技术实现，而是要清晰地阐明需要保护的各项条目。

制定网络安全策略是一件很复杂的事情，其主要复杂性在于网络安全策略必须能够覆盖数据在计算机网络系统中存储、传输和处理等各个环节，否则安全策略就不会有效。比如，保证数据在网络传输过程中的安全，并不能保证数据一定是安全的，因为该数据终究要存储到某台计算机上。如果该计算机上的操作系统等不具备相应的安全性，数据可能从那儿泄漏出去。因而，安全策略只有全方位地应用，才能是有效的。也就是说，该策略必须考虑数据的存储、传输、处理等。

由于制定合理的网络安全策略需要正确评估系统信息的价值，网络安全策略的制定并不是一件容易的事。（为了对数据进行有效的保护，）网络安全策略必须能够覆盖数据在计算机网络系统中存储、传输和处理等各个环节。

10.2　安全性指标

制定安全策略的复杂性还体现在必须决定哪个指标是最重要的，往往必须在安全性和实用性之间采取一个折衷的方案，例如，可以考虑：

数据完整性，即保护数据不被改变，也就是数据在发送前和到达后是否完全一样。
数据可用性，即在系统故障的情况下数据是否会丢失。
数据保密性，即数据是否会被非法窃取，也就是防止发生未经认可的访问。

10.3　安全责任和控制

许多组织发现他们无法设计一个安全策略，因为他们还没有明确信息控制的职责。这个问题通常可以从两方面来考虑：

帐户。考虑如何规定系统各用户对系统各项信息的访问权限，如何监督用户活动、记录用户活动情况等。

授权。对系统内每条信息，考虑如何规定各用户对它的操作权限，如只读、读写以及用户之间的权限转让等。

不管是帐户管理还是授权管理，关键问题是安全责任控制度，一个组织必须像管理有形资产如办公楼、机器设备一样对信息进行管理。

10.4 完整性机制

校验和与循环码校验技术可用于在数据意外破坏的情况下保证其完整性。使用这些技术，消息发送方可同时发送一个小的整数作为消息的检验值，接收方接到消息时只需重新计算一次检验值，并比较两检验值是否相同就可判断该消息是否正确了。

校验和以及循环码校验技术都不能绝对保证数据的完整性，有两个原因：第一，如果由于硬件故障使检验值和消息数据同时被破坏，则可能出现改变后的检验值和消息正巧匹配的情况；第二，如果恶意攻击导致数据被改变，攻击者可以为改变后的数据产生一个有效的检验值。

10.5 访问控制和口令

很多计算机系统采用口令机制来控制对系统资源的访问。每个用户都有一个秘密的口令。当用户需要访问被保护的资源时，就会被要求输入口令。

在传统的计算机系统中，简单的口令机制就能取得很好的效果，因为系统本身不会把口令泄漏出去。而在网络系统中，这样的口令就很容易被窃听。如果某用户通过网络传输口令到一台远程计算机上，在线窃听者就很容易获取该口令的副本。在线窃听在局域网上更容易实现，因为大多数局域网都是总线结构，任一台计算机都可以获得传输数据的副本。在这种情况下，就必须采取另外的保护措施。

10.6 加密与保密

为了保证在有线路窃听的情况下的数据保密性，必须对数据进行加密。加密的基本思想是打乱信息位元的排列方式，使得只有合法的接收方才能将其复原。其他任何人即使截取了该加密信息也无法解开。

目前存在几种加密技术。在一些加密技术中，消息发送方和消息接收方必须使用相同的密钥，该密钥必须保密。发送方用该密钥对待发消息进行加密，然后将其通过网络传输至接收方，接收方再使用相同的密钥对收到的消息进行解密。也就是说，消息发送方使用的加密函数有两个参数：密钥 K 和待加密消息 M，如加密后的消息为 E，则：

$$E=encrypt(K, M)$$

消息接收方使用解密函数把这一过程逆过来，就生成了原消息：

$$M=decrypt(K, E)$$

10.7 公共密钥加密

在很多加密方法中，为了避免危及安全，必须保持密钥是保密的。但有一种十分有趣的加密方法称为公共密钥加密法，它给每个用户分配两把密钥：一个称为私有密钥，是保密的；另一个称为公共密钥，随用户名一起公布，因而是众所周知的。该方法的加密函数必须具备如下数学特性：用公共密钥加密的消息除了使用相应的私有密钥外很难解密；同样，用私有密钥加密的消息除了使用相应的公共密钥外很难解密。

使用这两把密钥加密与解密的关系表示成数学形式如下：假设 M 表示一条消息，pub-u1 表示用户 1 的公共密钥，prv-u1 表示用户 1 的私有密钥，那么有：

$$M=decrypt\ (pub\text{-}u1,\ encrypt\ (prv\text{-}u1,\ M\ ))$$

和

$$M=decrypt\ (prv\text{-}u1,\ encrypt\ (pub\text{-}u1,\ M\ ))$$

公布公共密钥是安全的，因为这种方法的加密和解密函数具有单向性。也就是说，仅知道了公共密钥并不能伪造由相应私有密钥加密过的消息。

公共密钥加密法能够保证保密性。只要消息发送方使用消息接收方的公共密钥来加密待发消息，就只有消息接收方能够读懂该消息。因为要解密必须要知道接收方的私有密钥。因此，这种方法可以保证数据的保密性，因为只有接收方才能解密消息。

10.8 数字签名的鉴定

一种加密机制还可以用于验证消息发送方，这种技术称为数字签名。要在一条消息上签名，消息发送方使用只有接收方知道的密钥加密。接收方使用逆函数对消息解密。接收方知道是谁发送的消息，因为只有发送方拥有加密操作的密钥。为了保证加密的消息不被复制和重传，原始消息中可引入创建该消息的日期和时间。

公共密钥系统是如何实现数字签名的呢？过程是这样的，发送方通过使用自己的私有密钥对消息加密，实现签名；接收方通过使用发送方的公共密钥对消息解密，完成验证。原因是只有发送方知道私有密钥，所以只有他能发送可被公共密钥解密的密文。

有意思的是，如果采用双重加密的话，就可以使消息同时具有身份可验证性和保密性。所谓双重加密，是指首先使用发送方的私有密钥加密，再使用接收方的公共密钥对已加密消息进行再加密。其数学形式如下：

$$X=encrypt\ [pub\text{-}u2,\ encrypt\ (prv\text{-}u1,\ M\ )]$$

其中，M 表示原始消息，X 表示双重加密后的消息，*prv-u*1 表示消息发送方的私有密钥，*pub-u*2 表示消息接收方的公共密钥。

在接收端，解密过程是加密过程的逆过程。首先，消息接收方用它的私有密钥解除外层加密，这种解密只解除了一层加密，消息的数字签名仍未破解。然后，接收方使用消息发送方的公共密钥解除内层加密。这一过程表示如下：

$$M=decrypt\ [pub\text{-}u1,\ decrypt\ (prv\text{-}u2,\ X\ )]$$

其中，X 表示通过网络传送的加密消息，M 表示原始消息，*prv-u*2 表示消息接收方的私有密钥，*pub-u*1 表示消息发送方的公共密钥。

如果一条有意义的消息通过双重加密，那么它必定是保密、可信的。该消息必定到达指定的接收方，因为只有指定的消息接收方才拥有解除外层加密所需的密钥。同时该消息的身份一定是经过验证的，因为只有消息发送方才拥有必要的内层加密密钥。

10.9 包过滤

为了防止网络系统中每台计算机都可随意访问其他计算机以及系统中的各项服务，许多网站使用包过滤技术。如图 10-1 所示，包过滤器是在路由器中运行的一种程序，它由阻止数据包通过路由器在不同的网络之间穿越的软件组成。网络管理员可以配置包过滤器，以控制哪些包可以通过路由器，哪些包不可以。

包过滤器的工作是检查每个包的头部中的有关字段。网络管理员可以配置包过滤器，指定要检测哪些字段以及如何处理等等。比如，控制属于两个不同网络的计算机之间的通信，要检测每个包头部中的 source 和 destination 字段。上图中，要防止右边网络中 IP 地址为 192.5.48.27 的计算机和左边网络中的所有计算机，网络管理员配置包过滤器必须阻止所有 source 字段穿过路由器到达另一网络的包为 192.5.4827 的包通过。同样要防止左边网络中 IP 地址为 128.10.0.3 2 的计算机接收来自右边网络中的任意包，包过滤器必须阻止所有 destination 字段为 128.10.0.3 2 的包通过。

除了源地址和目的地址之外，包过滤器还能检查出包中使用的上层协议，从而知道该包所传递的数据属于哪一种服务。包过滤器的这种功能使得网络管理员能够对各种服务进行管理，比如可以过滤掉所有 WWW 服务的包而让电子邮件的包能得到较快的传输等。

网络管理员可以根据需要灵活配置包过滤器，以达到其所希望的过滤效果。通常，包过滤器的过滤条件是源地址、目的地址以及各种网络服务等的复杂的布尔表达式。凡是满足该过滤条件的包都会被过滤掉。比如，包过滤器可以同时过滤掉所有目的地址为 128.102.14 的 FTP 服务，所有源地址为 192.5.4 8.33 的 WWW 服务，以及所有源地址为 192.5 .48.34 的电子邮件服务。

10.10 互联网防火墙概念

包过滤器经常用来控制一个单位的计算机和内部网络与因特网之间的通信。如图 10-2 所示，包过滤器是连接内部网络和因特网的路由器的一部分。

用于保护一个单位的内部网络，使之不受来自外部因特网的非法访问的包过滤器，称为因特网防火墙。这个术语来源于两个结构间为防止火在它们之间蔓延而设置的物理防火边界。与传统的防火墙相似，因特网防火墙可用来防止因特网上出现的问题波及单位内部的计算机。

防火墙是互不信任的单位之间建立网络连接时最重要的安全工具。通过在网络联结的外围设置防火墙，组织可以界定一个安全边界，防止外界侵入组织内部的计算机。通过限制对一小部分计算机的访问，防火墙能阻止外界接触到组织内所有计算机、或防止大量有害数据引起内部网络瘫痪。

防火墙还可以降低系统成本。如果没有防火墙，外界可以向任意计算机传输数据包。因此，要保证内部网络的安全，必须使该网络内的所有计算机都安全才行。而有了防火墙，就可以通过它来灵活控制，使外界对内部网络的访问只能集中在几台计算机上，而系统中的其他计算机都不必提供安全措施，从而节省了一大笔开销。

第 11 章 数据库管理

11.1 概述

数据库（有时也拼写成 data base）也叫电子数据库，指的是任何数据或信息的总汇，也就是为了计算机快速搜索和检索而专门组织的信息。数据库的构建使各种数据处理操作中的数据的存储、检索、修改和删除变得容易。数据库可以存储在磁盘、磁带、光盘或一些其他副存储器上。

数据库由一个或一组文件组成。这些文件中的信息可以分解成多个记录，每个记录由一个或多个字段组成。字段是数据存储的基本单元，而且每个字段通常包括了由数据库所描述的实体的一个特点或属性信息。使用关键字和各种分类命令，用户可以快速搜索、重组、分组和选择许多记录中的字段，进行检索或创建特殊数据集合的报告。

数据库记录和文件必须组织成允许检索的信息。早期的系统按顺序整理（例如，字母顺序、数字顺序、或时间顺序）；直接访问存储设备的发展使通过索引随机访问数据成为可能。查询是用户检索数据库信息的主要方法。通常，用户提供一串字符，计算机按相应顺序搜索数据库，并提供字符串出现的源材料。例如，用户可以查讯一个人的姓是"史密斯"的字段的全部记录内容。

大型数据库的许多用户必须能够在任何时间内快速使用数据库中的信息。而且，大公司和其他机构倾向于建立很多包含相关数据、甚至重叠数据的文件，他们的数据处理经常要求链接几个文件中的数据。为了满足他们的要求，已开发了几种不同类型的数据库管理系统：非结构化数据库、层次型数据库、网络数据库、关系型数据库和面向对象的数据库。

在非结构化数据库中，根据实体的简单列表组织纪录，许多个人计算机的简单数据库是非结构化的。层次型数据库中的纪录是按树状结构组织的，记录的每一级分支成一套较小的分类。与层次型数据库不同，网络数据库通过设置链接或指针来建立不同记录集合之间的多重链接，而层次型数据库只为不同等级的记录集合提供单一链接。网络数据库的速度和通用性使它在商业中广泛应用。关系型数据库用于文件或记录的联系不能用链接表示的场合，一个简单的非结构化列表成为一个表格或"关系"，且多重关系可以通过数学关联来产生所需的信息。面向对象的数据库存储和处理更为复杂的数据结构，这种数据结构被称为"对象"，它们被组成了层次型的"类"，这些类可以继承链中更高级类的特性；这种数据库结构具有最好的灵活性和适应性。

许多数据库中的信息是由文档的自然语言文本组成的；面向数字的数据库主要由统计数据、表格数据、财务数据和原始科技数据之类的信息组成。小型数据库可以在个人计算机系统上维护，且可以由个人在家中使用。这些小型数据库和大型数据库在商业生活中变得越来越重要。典型的商业应用包括航空预定、生产管理、医院的医药纪录和保险公司的法律纪录。最大的数据库通常由政府机关、商业机构和大学来维护。这些数据库可以包括诸如摘要、报告、法规、线路服务、报纸杂志、百科全书和各种目录之类的文本。参考数据库包括参考书目或索引，可作为书、期刊和其他出版物的信息地址指南。目前有上千种可公开访问的数据库，覆盖了从法律、医药和工程到新闻时事、游戏、分类广告和教育课程各个主题。科学家、医生、律师、财务分析师、股票经纪人和各类研究人员越来越多地依靠这些数据库来快速、有选择性地来访问大量信息。

11.2 数据库模型

在数据库中构建数据的关系有多种不同方法。创建数据库的第一步是选择表示数据的模型，现有多种构件用于构造更复杂的数据库模型。

模型，经常被叫做模式，用来描述数据库的总体特性。就像一本书中的表格内容一样，数据库模型确定数据库的主要部分（如文件、记录和字段），并说明这些部分如何结合在一起。

数据库模型包括非结构化文件、关系型、层次型、网络型、面向对象型和文本。

11.2.1 非结构化文件

所有数据库模型中最简单的就是非结构化文件，也叫表格。非结构化文件是包含数据行（记录）和列（字段）的单一文件，它形成了一个二维电子表格。例如，假设你要为邮购业务产生一个客户文件，可以使用每一个客户一个记录的非结构化文件的模型，用客户的名字和地址来生成单个字段，并用一个独特的客户身份（ID）字段和这些数据相结合。ID 字段是解决客户名字完全相同问题的关键字段（如：西雅图的马克·史密斯和纽约的史密斯）。

虽然像客户文件这样的单一非结构文件可用于跟踪客户、准备邮寄列表，但它不能创建一个完整邮购程序。因此，需要一个不同的数据库模型（如表 11.1 所示）。

表 11-1

客户身份	名字	姓	地址	城市	街道	邮编
001	Smith	Mark	656 246th St.	Roslyn	NY	11576
002	Daolt	Shelia	12 Windsong Dr.	Arlington	VA	22201
003	Nasser	James	123 Watercress Ln.	Midvale	UT	84074
004	Smith	Mark	807 W 19th St.	Seattle	WA	98168

11.2.2 关系型数据库

1970 年，当时在 IBM 工作的 E. F. Codd 发表了一篇题为"大型共享数据库的关系模型"的论文。现在这篇论文被视为关系型数据库技术整个领域的开端。

关系型模型使用一种或多种非结构化文件或表格，并以每个表格中共同字段为基础建立表格之间的关系。每个非结构化文件或表格称为一个关系。例如，一个邮购应用必须使客户以及销售的产品井然有序，所以需要一个清单文件。清单文件描述销售的每个项目，它的关键字段是一个独特的项目数字。

邮购应用的其他组成是接受订单和记帐。为了接受订单，订单文件应有订单号、订单数据、客户号、项目号和订货数量。为了制成账单或发票，发票文件应有发票号、客户 ID、日期、订购项目和数量、价格，以及发票是否已经付出。

现在数据库由四个非结构化文件组成：客户、清单、订单和发票。注意相关数据中有一些冗余和重复。现在我们做好数据操作准备。

填写订单大致按下列程序操作：客户提交商品订单，然后为订单指定唯一的订单号。订单文件由带有发票文件数据的关联客户文件更新。类似地，以相关清单文件、订单文件和客户文件中的共同字段为基础制成发票。

在关系型数据库术语中，出现了应用于相互关系或文件的一套运算符号，这就是众所周知的关系代数。运算符号从现存的文件里提取数据，然后产生期待的结果。常见的运算符号是连接符和项目运算符，前者用一个公共字段把两个分开的文件结合在一起，后者从现存文件中选择字段生成新的文件。

使用关系模型的产品例子是 IBM 的 DB2 数据库，Oracle 公司的 Oracle 数据库和 Sybase 公司的 Sybase 数据库管理系统。

11.2.3 层次型

层次模型比关系模型老得多。它把结构化数据构成一颗倒置的树，创建数据之间的关系。其记录包括：

（1）单一根或主键字段，用于确认类型、地址或记录的顺序。

（2）下属字段的变量数，用于定义记录内的其余数据。

因为在商业应用中随处可见层次关系而开发了层次模型。众所周知，组织图经常用于描述层次关系：顶级管理处于最高层，中间管理处于较低层，而操作雇员处于最底层。注意，在严格的层次下，每个管理层可以有许多雇员或雇员层，但每个雇员只有一个经理。层次数据的特点是数据中的这种一对多关系。

作为另一个例子，考虑一个简化的飞机零件数据库。同大多数系统一样，飞机是由一系列部件组成的，部件是由子部件组成的，等等。使用层次方法，可以建立如下记录与字段之间的关系：最高级包括主要部件，如机翼、机身和座舱，第二级包括每个主要部件的子部件；再下层包括特殊零件数和零件信息。

这种方法可以很方便地答复客户关于零件和零件利用率方面的查询，但不方便查询零件在什么飞机上。在获得有关特殊零件和飞机信息前，必须首先检索每个主要部件，而且必须经过几个层次来获得零件信息。

在层次方法中，当创建数据库时必须明确定义每个关系。层次数据库中每个记录只允许有一个关键字段，且任意两个字段间只允许有一种关系。因为实际数据并不是总符合这样一个严格的层次关系，这就可能产生一个问题。例如，机翼上的铆钉可能与机身铆钉完全相同。

使用层次模型的商业化数据库产品有 IBM 的 IMS 和 Cullinet 的 IDMS。

11.2.4 其他数据库类型

我们已经提到过文本模型，还有其他一些数据库类型值得一提。

在连接表结构中，下级记录可以与不只一个父记录链接，网络模型通过连接表结构建立数据之间的关系。这种方法把记录与链接结合起来，其链接叫做指针。指针是指示记录位置的地址。有了网络方法，下级记录可以与关键记录链接，且同时它自身作为关键记录与其他下级记录相链接。历史上网络模型已表现出优于其他数据库模型的优势。今天，这种特性只在高容量、高速度传输过程中仍为重要，如，自动柜员机网络或航空预定系统。

面向对象的模型使数据组合成代表某种对象的集合。例如，与飞机机翼有关的所有数据。更重要的是，面向对象的模型也允许记录从前辈记录继承信息。例如，在邮购应用程序中，鞋的订单将从订单文件中继承诸如客户名称和地址之类的信息。

11.3 数据挖掘

一个迅速扩大且与数据库技术密切相关的主题是数据挖掘，主要包括数据集合中的发现模式技术。数据挖掘在很多领域都已经成为一个重要的工具，包括市场营销，库存管理，质量控制，贷款风险管理，欺诈检测和投资分析。数据挖掘技术甚至应用在一些看起来不太可能的场合，例如在识别特定基因编码的 DNA 分子的功能和表征生物性质的应用中。

数据挖掘活动与传统的数据库询问的区别在于数据挖掘的目的是识别以前未知的模式，而不是传统的数据库查询，仅仅请求对存储事实的检索。此外，数据挖掘更适用于静态数据的集合，即数据仓库，而不适用于受频繁更新影响的"在线"业务数据库。这些仓库通常是数据库或数据库集合"快照"。它们被用来代替实际业务数据库，因为寻找模式在静态系统比在动态系统中更容易实现。

我们还应该注意到，数据挖掘的对象不仅限于计算的领域，它已经延伸到统计数据中。事实上，许多人会认为，既然数据挖掘是源于尝试进行大型统计分析和不同的数据集，那么它应该是统计数据的应用程序，而不属于计算机科学领域。

数据挖掘的两种常见的形式是类的描述和类的区别。类的描述处理的是识别表征一组给定数据项目的属性，而类的区别处理的是识别两组数据的属性。举个例子，类描述技术被用来识别购买小型经济车的人的特征，而类区别技术被用于找出能够区别从购买新车的客户手中购买二手车的客户的属性。

数据挖掘的另一种形式是聚类分析，其目的是发现类。请注意，这不同于类的描述，其目的是在已经确定类的成员中发现其属性。更确切地说，聚类分析试图找到导致发现分组的数据项的属性。例如，在分析观看特定电影的人的年龄信息时，聚类分析可以发现，客户群分解成两个年龄组，一个 4 至 10 岁年龄组和 25 至 40 岁年龄组。（也许是电影吸引了孩子们和他们的父母？）数据挖掘还有一种形式是关联分析，其涉及寻找数据组之间的关系。关联分析可以揭示出买薯片的顾客也买啤酒和汽水，或者在传统工作日购物的人也会有退休福利。

孤立点分析是数据挖掘的另一种形式。它试图确定不符合规范的数据条目。孤立点分析可用于识别数据集合中的错误，可以在客户正常的购买模式中检测到突然的偏离以识别出信用卡被盗刷，或许也可以通过辨识异常行为来确定潜在的恐怖分子。

最后，还有一种称为序列模式分析的数据挖掘形式，能够确定随着时间的推移的行为模式。例如，序列模式分析可能会揭示经济系统的趋势，如股市或环境系统，如气候条件。

正如我们的最后一个例子，数据挖掘结果可以被用来预测未来的行为。如果一个实体拥有表征一类的属性，则该实体可能会表现得像是类的成员。然而，许多数据挖掘项目仅仅旨在更好地了解数据，人们见证了在解开 DNA 的奥秘过程中利用到了数据挖掘。在任何情况下，数据挖掘应用的范围可能是巨大的，因此，在未来几年数据挖掘有望成为一个活跃的研究领域。

需要注意的是数据库技术和数据挖掘是近亲，因此对其中一个的研究将对另一个也产生影响。数据库技术被广泛地用于给数据仓库赋予以数据立方体（从多个角度查看数据——立方体一词用于推测多个维度的图像）的形式代表数据的能力，而该数据立方体使得数据挖掘成为可能。数据挖掘的研究人员改进执行数据立方体的过程将反过来为数据库设计的领域提供帮助。

最后，我们应该认识到，成功的挖掘比数据集合模式识别包含的更多。智能判断必须应用于确定这些模式是否有意义或仅仅巧合。某个特定的便利店已经售出了大量中奖彩票的事实，对一个打算买彩票的人来说可能不应该被认为是有意义的，但发现买快餐食品的顾客也更倾向于购买冷冻食品，这对杂货店经理来说可能就是有意义的信息。同样，数据挖掘包含广泛的道德问题，涉及到代表数据仓库的个人权利、结论的准确性和利用、甚至还有摆在首位的数据挖掘的适当性。

第 12 章　多媒体和计算机动画

12.1　多媒体

多媒体是文本、声音、图画、动画和视频等信息组合的表现形式。普通的多媒体计算机应用程序包括游戏、学习软件和参考资料。大多数多媒体应用程序包括事先定义的关联，称为超级链接，使用户能够在媒体元和主题之间进行切换。

周到仔细地表现媒体能增强表现效果，在形式上就像人脑漫无边际的联想。超级链接提供的连通性将多媒体从带有图形和声音的静态显示转变成变化无穷的、知识丰富的交互体验。

多媒体应用程序是计算机程序，它们通常存储在光盘上。它们也可以驻留在万维网上，万维网是因特网（国际通讯网络）媒体丰富的组成部分。万维网上提供的多媒体文档叫做网页。由超级链接将信息链接起来后通过特殊的计算机程序或计算机完成的。用来建立网页的计算机语言叫做超文本链接标示语言（HTML）。

多媒体应用程序通常要比只用文本表示要求的信息需要更大的计算机内存和更高的处理能力。例如，运行多媒体的计算机必须有快速的中央处理器（CPU），它是为计算机提供计算能力和控制的电子电路系统。多媒体计算机还需要额外的电子内存来帮助 CPU 进行计算，使视屏能画出复杂的图像。计算机还需要高容量硬盘来存储和检索多媒体信息，还需要一个光盘驱动器来播放 CDROM 应用程序。最后，多媒体计算机必须有一个键盘和一个指示设备，如鼠标或轨迹球，这样用户可以引导多媒体元件之间的关联。

12.1.1　视频元

图像越大、分辨率越高、色彩越丰富，在计算机屏幕上就越难显示和操作。照片、图画和其他静止图像必须转换成计算机能操作和显示的格式。这样的格式包括位图图形和矢量图形。

位图图形用小点的行和列来表示图像。在位图图形中，每个点都用行和列准确地描述其位置，很像在一座城市中每个房子都有一个准确的地址。最常见的位图格式有可交换图像文件格式（GIF）、标签图像文件格式（TIFF）和 Windows 位图（BMP）。

矢量图形是用数学公式来重现原始图像。在矢量图形中，点不是用行和列的地址来定义的；相反，它们是由相互之间的空间关系定义的。因为它们的组成点不严格限制在特定的行和列上，矢量图形可以更容易地复制出图像，而且它们在大多数视屏和打印机上看上去效果更好。常见的矢量图形格式有附录显示格式（EPS）、Windows 图元文件格式（WMF）、惠普图形语言（HPGL）和麦金托什图形文件格式。

视频元的建立、格式化和编辑要求由特殊的计算机部件和程序完成。视频文件可能很大，所以通常用压缩的方法来缩小，压缩是一种识别再现信息集的技术，如一行中有一百个黑点，然后再用一段信息来代替它，达到节省计算机存储系统空间的目的。

常见的视频压缩格式有多媒体文件格式（AVI），Quiktime 格式和运动图像专家组（MPEG）格式。这些格式能把视频文件缩小百分之九十五，但它们也会引起不同程度的图像模糊。

动画也可以包括在多媒体应用程序中，以便把运动加入到图像上。动画对模拟真实世界的情况特别有用，如喷气飞机的飞行。通过加入特殊效果，动画也能加强现有的图形和视频元，如将一个图像天衣无缝地拟合，合成到另一个图像上。

12.1.2 声音元

像可视元件一样，声音必须记录和格式化，这样计算机能在演示中理解和使用它。两种普通的音频格式是波形（WAV）声音资源文件和乐器数字界面（MIDI）。WAV 文件存储实际声音，很像音乐 CD 和磁带。WAV 文件可以很大，也许需要压缩，MIDI 文件不存储实际声音，而是能让混音器再现声音或音乐的指令。MIDI 文件比 WAV 文件小得多，但再现的声音不太好。

12.1.3 组织元

演示中的多媒体元要求有一个框架结构来赞助用户学习和与信息交互。交互元素包括弹出菜单、出现在计算机屏幕上带命令列表的小窗口或可供用户选择的多媒体元素。滚动条通常处于计算机屏幕一侧，使用户能够从大文档或图片的一部分移动到另一部分上。

通过超级链接可以增强多媒体表现元素的集成。超级链接创造性地使用彩色文本、或带下划线文本、或被称为图标的小图形连接表现多媒体的不同元素，用户可以把光标指向图标并点击鼠标进行操作。例如，一篇关于约翰·肯尼迪总统的文章可能包含他被暗杀的段落，一个超级链接就设在"肯尼迪葬礼"这个词上。用户点击带有超级链接的文本，就会被带到肯尼迪葬礼的视频演示上。这段伴有插图说明的视频内嵌超级链接，会把用户带到不同文化的葬礼演示，并配有各种丧葬歌曲。这些歌曲具有指向乐器演示的超级链接。这个超级链接链可以把用户带到他们在其他情况下不可能遇到的信息。

12.1.4 多媒体应用程序

多媒体对教育产生了巨大的冲击。例如，医学院用多媒体模拟手术能让未来的外科医生在计算机生成的"虚拟"病人身上进行手术。类似地，工科院校的学生用电路设计交互多媒体演示来学习电子基础，并立即实现、测试和操作他们在计算机上设计的电路。即使在小学里，学生使用简单但功能强大的多媒体写作工具来创作多媒体演示，提高报告和短文水平。

多媒体也用于商业应用。例如，一些娱乐场所提供多媒体游戏来让游戏者玩 Indy 赛车或在逼真的大型机器人座舱中互相打斗；建筑师用多媒体演示方法让客户浏览还没建成的房屋；邮购业务提供多媒体分类目录，使潜在购买者能够浏览虚拟展室。

12.2 计算机动画

计算机生成动画的代表性应用有娱乐（电影和动漫）、广告、科学和工程研究以及培训和教学。尽管我们在考虑动画时暗指对象的移动，但术语"计算机动画"通常指场景中任何随时间而发生的视觉变化。除了通过平移、旋转来改变对象的位置外，计算机生成的动画还可以随时间进展而改变对象大小、颜色、透明性和表面纹理等。广告动画经常把一个对象形体变成另一个。例如，将一个汽车油罐变成汽车发动机。计算机动画还可以通过改变照相机的参数，例如位置、方向和焦距。我们还可以通过改变光照效果和其他参数以及照明和绘制过程来生成计算机动画。

许多计算机动画的应用时要求显示真实感。利用数值模型来描述的雷暴雨或其他自然现象的精确表示对评价该模型的可靠性是很重要的。同样，培训飞机驾驶员和大型设备操作员的模拟器必须生成环境的精确表示。另一方面，娱乐和广告应用有时较为关心视觉效果。因此可能使用夸张的形体和非真实感的运动和变换来显示场景。但确实有许多娱乐和广告应用要求计算机生成场景的精确表示。在有些科学和工程研究中，真实感并不是一个目标。例如，物理量经常使用随时间而变化的伪彩色或抽象形体来显示，以帮助研究人员理解物理过程的本质。

12.2.1 动画序列的设计

通常，一个动画序列按照以下几步进行设计：

- 故事情节拆分
- 对象定义
- 关键帧描述
- 插值帧的生成

这种制作动画片的标准方法也适用于其他动画应用，尽管有许多专门的应用并不按此序列进行处理。例如，飞行模拟器生成的实时计算机动画按飞机控制器上的动作来显示动画序列。而可视化应用则由数值模型的结果来生成。对于逐帧动画，场景中每一帧是单独生成和存储的。然后，这些帧可以记录在胶片上或以"实时回放"模式连贯地显示出来。

剧本是动作的轮廓。它将动画序列定义为一组要发生的基本事件。依赖于要生成的动画类型，剧本可能包含一组粗略的草图或运动的一系列基本思路。

为动作的每一个参加者给出对象定义。对象可能使用基本形体如多边形或样条曲线进行定义。另外，每一对象的相关运动则根据形体而指定。

一个关键帧是动画序列中特定时刻的一个场景的详细图示。在每一个关键帧中，每一个对象的位置依赖于该帧的时刻。选择某些关键帧作为行为的极端位置。另一些则以不太大的时间间隔进行安排。对于复杂的运动，要比简单的缓慢变化运动安排更多的关键帧。

插值帧是关键帧之间过渡的帧。插值帧的数量取决于用来显示动画的介质。电影胶片要求每秒24帧，而图形终端按每秒30到60帧来刷新。一般情况下，运动的时间间隔设定为每一对关键帧之间有3~5个插值帧。依赖于为运动指定的速度，有些关键帧可重复使用。一分钟没有重复的电影胶片需要1440帧。如果每两个关键帧之间有5个插值帧，则需要288幅关键帧。如果运动并不是很复杂，我们可以将关键帧安排得稀一点。

可能还要求其他一些依赖于应用的任务。包括运动的验证、编辑和声带的生成与同步。生成一般动画的许多功能现在都由计算机来完成。

12.2.2 通用计算机动画功能

开发动画序列中的某几步工作很适合计算机进行处理。其中包括对象管理和绘制、照相机运动和生成插值帧。动画软件包，如Wavefront，提供了设计动画和处理单个对象的专门功能。

动画软件包中有存储和管理对象数据库的功能。对象形状及其参数存于数据库中并可更新。其他的对象功能包括运动的生成和对象绘制。运动可依赖指定的约束，使用二维或三维变换而生成。然后可使用标准函数来识别可见曲面并应用绘制算法。

另一种典型功能是模拟照相机的运动，标准的运动有拉镜头、摇镜头和倾斜。最后，给出对关键帧的描述，然后自动生成插值帧。

第 13 章　物联网的剖析

设计一个新的物联网的网络体系结构可能看上去是一项令人怯步的任务。但是，确实需要这样一个全新的方案。物联网的环境是如此的不同，并且连接其中的设备也是各种各样的，因此，自从互联网的原型建立以来，一个前所未有的网络挑战就摆在了我们面前。

在建立发展物联网的过程中，传统互联网的一些经验教训和其他的变换技术为我们提供了一些基本准则：

- 指定的内容越少越好，为其他人的创新留下足够空间
- 系统必须设计成为不完美的：不是寻求消除错误，而是适应和调解这些错误
- 分度的联网功能和复杂度仅在需要时应用
- 系统结构应由一些简单的概念发展而来，而这些概念使用自然现象中的模拟量来建立一个完整的系统
- 其意义可以从数据中实时获取

通过减少网络知识和网络边缘资源的方法，正在出现的这种物联网结构被规定为包含大量的市场参与者的形式。同时，这种结构必须具有极好的容错性，而且在这一层次上可以保持时断时续的连接。（与直观相反，最好的方式是在网络边缘简化协议，而非使之变得更为复杂。）

接下来，增加网络能力的复杂程度被应用于传统互联网的网关，因为在传统互联网中，传播节点可以为军方非精细设备提供通讯服务。

最后，通过对积分函数中大量数据的提取可以获得其意义，而积分函数提供了物联网的人机界面。这一层的监管只能由网络的最高层来负责；而一些简单的设备，就像蜂巢里的工蜂一样，不需要承载大量的计算工作或是占用较多的网络资源。

为了探索这种新结构的所需，首先要做的就是摒弃当前的网络状态和形式。

13.1　传统的互联网协议不适用于大多数的物联网

当我们在考虑物联网如何运转时，忘记传统互联网规划的世俗认知是大有裨益的——尤其是广域网（WAN）和无线网的相关概念。在传统的广域网和无线网络中，其带宽和频谱是昂贵且有限的，并且需要传播的数据往往是大量的且在不断增长的。尽管在连接台式机（至大多数传统的互联网）时预留超量的数据通路是司空见惯的，但在广域网和无线网中通常是不实用的——因为花费过于昂贵。由于传输媒介占据了大量的成本负担，并将这种负担转嫁到了用户身上，无线网络的成本是同等使用 IP 的有线网络造价的十倍。

除成本外，还存在潜在的数据丢失和（在无线网络中的）冲突问题。传统的网络协议包含大量对信息完整性的检查和确认，从而尽可能减少昂贵的重复发送。这些限制就形成了如今熟悉的协议栈，如 TCP/IP 和 802.11。

13.1.1　"啁啾"的介绍

然而，在大多数的物联网中，情况变得完全不一样了。可以确定的是，无线网络和广域带宽的造价依旧很高。而且由于多数网络边缘的连接——物联网前沿，可以说是——无线连接或者是有失真的，这也是任何物联网结构都必须考虑的因素。但是大部分设备的数据总量是极低的，而且任何单个信息的传递是完全无需检验的。正如之前所讨论的，物联网是有失真的且断续的，所以终端设备必须被设计为可以在短时误发送或

接收情况下也可以正常工作的，甚至是长时间的情况也可以正常工作。正如之前所提到的，这种自足性能消除了任何单个信息的临界性。

在彻底考虑物联网的全部所需，回顾了所有存在的可能之后，很显然我们要定义一种新的数据帧或包。这种新的数据包只提供网络边缘简单物联网设备开销和功能的数量——仅此而已。这些被称为啁啾的小数据包是物联网新兴结构的基本结构单元。在很多方面，啁啾和传统的互联网协议包都是不同的。啁啾的基本特性如下：

- 啁啾只包含最小的负载开销，"箭头"的传递，简单的非唯一地址，适度的校验
- 啁啾被设计为固有的个体非临界性
- 因此，啁啾不包含中继和回执协议

对于载有啁啾的传统互联网来说，任何所必须的额外的功能，如全局寻址，路由等，都被其他网络设备通过对接收取简单啁啾增加信息的方式所自动处理。因此，啁啾包里没有关于这些功能的规定。

轻巧的和一次性的啁啾

与传统的网络数据包结构形成对比的是，物联网啁啾数据包像花粉或者鸟鸣一样，具有如下特点：轻巧的，可被广泛传播的，只对"感兴趣的"积分函数或终端设备有意义。物联网是以接收方为中心，而不是发送方，IP 则反之。由于物联网啁啾具有轻巧和个体非临界性的特性，所以其在重试的关注极为有限，而这样的重试经常会导致广播风暴，在 IP 里是非常危险的。

的确，有效的物联网传播节点将修剪和打包广播（如图 13-1），但是源于终端设备的周期性或插话性的广播风暴则是一个小的多的问题，原因是啁啾轻巧的特性（导致较少的拥挤）和个体非临界性。因此，过多的啁啾可能会在需要的情况下被传播节点丢弃。

13.1.2 物联网的功能性需要

考虑到基于极小元素组成的大量网络，这种全新的网络观点意味着巨大的数据包，发布者的安全保证和任何单个信息的确定传递都不是必须的。在某种意义上，相比于"男性化"IP 结构（面向发送者），这使得物联网更"女性化"（面向接收者）。

但是，如果没有任何数据传输的话，这样的物联网是没有意义的。如何管理这些具有已知不可预测性的连接呢？或许有些出人意料，答案是预留超量的空间——但前提是仅限于啁啾设备和传播节点之间。也就是说，这些短小简单的啁啾可能会被一遍遍地发送，作为一种强制手段来确保部分啁啾得以通过。

13.1.3 冗余的效率

如图 13-2 所示，由于这些数据块极小，这种在物联网边缘预留超量空间的成本也是极小的。（通常是由本地的 Wi-Fi，蓝牙，红外等进行处理，因此它们不被任何载体所计量。）因此，这种方案的好处是巨大的。因为没有任何个体信息是临界的，所以不会发生错误恢复或整体性检测的开销（为避免混乱消息而进行的基本校验和除外）。每一个啁啾信息只包含一个地址，一个简短的数据字段，还有一个校验和。在某种程度上，这些信息就是 IP 数据报所包含的内容。在很多方面，啁啾和简单网络管理协议（SNMP）也很相似，即简单"获取"和"设置"的功能特性。

重要的是，终端设备合并啁啾信息负担的成本和复杂程度将会是极低的——因为这些都由物联网本身承载。最有效的集成方案可能是"在芯片上啁啾"，这样可以在一块简单标准化的芯片上实现最小的数据输入/输出和传输/接收功能。

如前文所述，啁啾也包含了"箭头"的信息，即定义了信息的总体方向：要么是指向终端设备，要么是指向积分函数（如图 13-3 所示）。流向或流出终端设备的信息只包含终端设备的地址；其指向哪儿或者是从哪儿来对大多数的简单终端设备来说都是不重要的。这些设备只是起广播或者监听作用，而本地的相关性才

是起关键作用的。

因此，终端设备可能会淹没在无尽的数据流中。它们可能不停地广播并且信任传播节点和积分函数，而在网络的别处则会删除和忽略冗余信息。同样地，它们也会在检测到一条变化的信息并对其做相应的动作前接收到无数相同的信息。

本质上来说，从本地反复发送信息的角度来看，啁啾协议是"不经济的"，而本地的带宽是非常便宜甚至是免费的（实质上是"离线的"）。但是传播节点被设计来减小多余或者重复的转发流量，因此广域网成本和流向传统互联网的流量都被大幅的减小了。

要注意的是，与传统网络终端设备（如智能手机和笔记本电脑）不同的是，绝大多数的物联网终端设备很可能不同时具有发送和接收功能（如图 13-4 所示）。举个例子，如空气质量传感器只需要发送它所监测的某种化学物质的当前状态即可。开关打开后，它开始不停地传送啁啾信息，直到开关关闭为止。这将极大地简化绝大部分的终端硬件及其所需的嵌入式软件。

13.2 这些都是相对而言的

啁啾数据包的详细结构不在本章讨论，但是一个简短的介绍在这里是有必要的。物联网数据包和其他数据包在形式上的关键不同在于其包内变量的含义都是相对的。也就是说，（这里用 IPv6 举例）对于包定位头，地址等都没有一个固定的定义。

如图 13-6 所示，标号"-"用来代替固定形式的定义，使接收设备可以将信息定义为发送地址，传感器和数据类型，传输箭头等。这些标号既是公用类型也是专用类型。

公用类型标记，在每一个物联网数据包里都存在，其使得接收设备能够"解析"入局通信量。当一个公共类型标记被识别出来，接收设备检测标记前后的数据，检测这些特定的位是为了决定其余的数据包将被转发和/或采取行动。接收设备通常情况下无须对数据包进行检测，除非是指定的区域位置并且观测到公共类型标记。公共类型标记包括前文提到的基本传输箭头，用于数据包检测的四位校验和等。数据区间的位不属于路由，且验证信息仅被作为本层校验一个数据负载。

13.2.1 格式的灵活性

物联网啁啾数据包中公共类型标记的存在使得物联网数据包的长度可以根据具体应用、设备类型或消息模式进行变化。带有不同数量的公共数据字段的物联网数据包被定为不同的类型，这样可以允许需要附加环境的应用加载足量的信息，也可以允许大部分基本设备和通用物联网数据包传播获得最小的开销。

公共类型标记的灵感来源于自然界，包括转录和读取编码在基因 DNA 里的遗传信息，而这些信息能够产生生命环境所必须的蛋白质。DNA 螺旋链也会包含一些重复的内容和不被读取的"垃圾"部分，而螺旋链上的标记就是用来指示转录"开始"和"停止"的位置。接收设备采用相同的方式借助公共类型标记来检验物联网啁啾数据包，而不需要特定的字节数或其他开销产生的限制。

13.2.2 专用类型标记的自定义和可扩展性

专业类型标记被允许存在于公用类型标记所定义的通用"数据"区间内，从而使得其可以根据特定的应用和制造商等自定义数据形式。接收设备可以通过同时借助公共和专用类型标记来解析数据流的办法来为特定需求定位信息。

13.2.3 寻址和"节奏"

如前所述，数十亿物联网终端设备将非常便宜，且可能是由世界各地的生产商制造的，而多数的生产商将不具备广泛的网络知识。为此，想要确保数千亿物联网终端设备地址集中数据库中每个地址的唯一性几乎是一件不可能完成的事情。

物联网啁啾数据包中部分公共信息将会是一个简单的、非唯一的四位设备 ID，该设备 ID 通过印刷电路板布线、硬件带、指拨开关或其他类似的方法进行应用。如第 8 章中所述，设备 ID 将与一个随机产生的 4 比特模式相结合来降低连接在同一个本地传播节点的两个终端设备具有相同标识的可能性。（这种比特的组合也被用来改变无线网络环境的传播速率，从而避免"死锁"现象的产生。）

如果特定的应用需要额外寻址的专一性和/或安全性，可以把这些信息加载到物联网数据包的专用空间里。

13.2.4 种类

包含在物联网啁啾数据包里最终的公共信息是 256 种可能的啁啾"种类"的其中一种。如第 8 章所述，这些啁啾种类先按照类型和应用线路被区分，例如不同类型的传感器、调节阀、红/黄/绿状态指示器等。这些啁啾的种类按照通用到具体的方式去定义，并足够广泛且可扩展以保证任何类型的物联网应用。正如上述，对一些需要更细分类特定应用或设备，这些自定义的信息可以在数据区间中的专用类型标记间定义。

对啁啾数据包的分类对物联网有着最为深远的意义：使数据分析程序具有发现和扩充新数据源的能力，而这样的能力是基于关联性信息社区的。由于这些类型和分类信息是"外部的"，它们可以被很多物联网元素识别和执行，例如积分函数和传播节点（如果装备了相关的发布代理，则也会一起动作）。

这样，监视管道上压力传感器的积分函数可以找到附件的温度传感器来寻找它们之间的关系以提供更为丰富的信息。单是啁啾数据包的种类和区分就传递了一些潜在的信息，并结合其他信息将这些信息分析和定位，这一过程是啁啾数据流向前传递的过程中在整个网络进行着的。

即使变送传感器因不同应用、不同机构或不同时刻而安装，物联网的这一特性也是千真万确的。不同"公共"广告类型和分类的选择使得啁啾数据流被更广泛的使用（和再利用），这一点是借助物联网随时间"学习"到而建立的动态发布/订阅关系而实现的。

这样做的好处是无需增加终端设备的负担。顾名思义，大部分的物联网终端设备都是及其简单的，这些被设计用来接收物联网啁啾数据包的终端设备只需要处理协议中最基本的元素（举个例子，利用公共类型标记去识别出指向这些设备本身的数据包并读出这些数据）。物联网元素更加广泛地利用了啁啾数据包的性能，这些元素必须传递或分析来自多个终端设备的数据，尤其是传播节点和积分函数。

13.3 将网络智能用于传播节点

如前所述，即使采用如此高效的啁啾协议，混乱的物联网数据流很明显也会堵塞网络，因此单个终端设备以上的层次必须采用更为智能的方法。这也就是传播节点的责任，它们是构建整体网络拓扑结构的设备，负责组织构成物联网的设备与设备之间的相互关系。

典型的传播节点是硬件和软件的结合，类似于 WiFi 的接入点。它们"控制"着本地的终端设备，这点意味着它们实际上可以和这些终端设备进行交互，这种交互通常发生在传播节点的无线传输范围内。它们可以专门用来从各种各样的终端设备上接收啁啾数据包。最终，在像拉斯维加斯一样大的城市内，可能会有数以万计或十万计的传播节点。传播节点会利用它们掌握的邻接信息去形成一幅网络的近程图像。它们会确定范围内附近的传播节点，以及和这些传播节点直接或间接连接的终端设备和积分函数。这些信息被用来形成网络拓扑结构：消除环路和创造已存在的代替路径。

在广播到其相邻节点之前，传播节点会智能地打包和修剪各种啁啾信息。在检验公共类型标记、简单校验和传输的"箭头"（指向终端设备或者积分函数）之后，损坏或者冗余的信息会被丢弃。通过一个相邻节点传播的几组信息会被打包成一个"中继"信息——一个小的数据"流"，为的是更有效的传输。接收到的"中继"信息会被解压和重新压缩。

一些类型的传播节点会含有一个软件发布代理。这样的发布代理与特定的积分函数交互来优化代表积分函数的转发数据。带有发布代理的传播节点可能会"偏向于"转发基于路由指令的特定方向的一类信息，该指令是由偏爱与特定的功能、时间或地理上"邻近"的终端设备通信的积分函数发出的。在物联网终端设备的邻域内，积分函数将主宰所有的通信流来根据需要获取数据或者修改参数。

为了发现新的终端设备，传播节点和积分函数又会变得和传统网络结构类似。当一个信息来自或者要发送到新设备时，传播节点会转发这些信息并在报表上添加地址（如图 13-7 所示）。如果有设备离线或者被移动，适当的超时算法将允许修剪相邻性的报表。

13.3.1 传输和功能结构

新兴的物联网结构包括两个完全独立的网络拓扑结构：传输和功能，如图 13-8 所示。其中，由传播节点（和全球互联网）构成的传输结构是基本构架，在其之上所有的信息流可以自由移动。功能结构是独立于物理通路的虚拟的相关"区间"或"邻域"，它由积分函数所建立。

物联网的传输网络部分的允许过程中，几乎不包含具有实际意义的数据啁啾。如前所述，传播节点建立的传输网络是基于更为传统的网络概念和路由算法。终端啁啾设备可以通过多种方式连接到传播节点：广播或光波的无线方式、电力线网络和直接的物理连接等。单独的传播节点可以连接大量的啁啾设备并为之提供服务。除非是积分函数更倾向于该传播节点，基本的模型是"混乱转发的"。

传播节点在传输需要的情况下会捆绑和转换啁啾数据流到相邻的数据节点，并从那里转发到积分函数或者啁啾设备。传播节点间的连接是典型的传统网络协议，例如 TCP/IP，但是它也可以是基于啁啾的。

除了传输及其简单的啁啾外，由传播节点建立的更高层协议报文还包括额外的背景信息，这点在啁啾里是没有的。这些数据可能包括和位置、时间或其他因素有关的额外信息，如图 13-9 所示。因此，在不增加大量终端设备和网络成本及复杂性的前提下，传播节点增进了啁啾数据流的利用率。这样的背景信息只能被传播节点添加，并由积分函数所解析。

物联网传输结构和许多传统网络结构的重要区别是根本上的平等，这点类似于气流中携带的各种花粉。在"信任""通信"和"控制"因素的约束下，传播节点将转发来自任何终端设备或者积分函数的物联网信息流。那么物联网就可以"搭载"于现有的架构上，并且每个新的传播节点都可以为各类用户和积分函数增加功能。根本上来说，传输网络拓扑结构不能建立（或限制）物联网的功能网络拓扑结构，其是由积分函数建立的。

13.3.2 功能网络拓扑结构

在说明了传输网络结构（如前文所述）为啁啾数据包提供双向（"向下"传给啁啾设备，"向上"传给积分函数）转发服务后，注意力应该转移到物联网功能结构上，其覆盖于传输结构之上，在某种程度上类似于花粉的传播也弥漫在整个空气当中一样。

那么，物联网的功能网络结构的问题就不是（物理的或者虚拟的）"线"如何连接的问题，而是如何确定相关信息的问题。新兴的互联网结构根本上来说是一个由积分函数驱动的"发布和订阅"的模型。这也是面向接收者的，由处在传输"箭头"远端的机器决定哪些数据是相关和有用的。

13.3.3 由积分函数所定义

此时，应当对积分函数进行一个简要的介绍了。积分函数可能有各种各样的物理形式，而某个和传统互联网（可能通过一个过滤网关）有唯一连接的机器也可能配备了多种逻辑积分函数。从功能上来说，它们在某种程度上是一组选定端点关系的自主创建者。

举个例子，假设有一个积分函数被设计用来监测某农业企业一个偏远场地的含水量（如图 13-10 所示）。水分传感终端设备每隔一段时间广播一次指示周围土壤含水量的啁啾信息。这些小的啁啾数据包含一个指向积分函数的传输"箭头"。

农业企业（或者其他任何人）在一定范围内配置的传播节点，用来接收啁啾数据。如前所述，这些啁啾信息离开传播节点时被捆绑了额外的背景信息，例如一个完整的 IPv6 地址和位置信息，这样可以对某个特定的传感器有着更为准确的位置信息和辨识度，而这是简单啁啾无法做到的。传播节点的传输网络实质上是通过传统的互联网把这些数据流"发布"出去的。

13.3.4 从物联网收集信息

之前的表述说明，在某个单独的农业供应商安装了自己的终端设备传感器传播节点的情况下，一个虚拟的专用传感器网络利用传统的互联网建立路由路径，之后根据需要对网络进行监控。当然，很多物联网大数据"邻域"都是以这种方式创建的。但是对建立网络来说，还存在其他巨大的可能，如一个依赖于物联网元素提供数据的网络，而这些元素不是被单个来源所拥有、管理和控制的。

在西方世界新兴的社会网络文化里，群众外包和数据共享变得越来越普遍。鉴于此，个人和组织会选择安装传感器、摄像机和其他的本地设备，借助这些设备公开地提供物联网数据流（值得注意的是，现在很多个人和团体都已经这么做了，他们借助网络摄像头、空气传感器，并使用传统的互联网协议，如 IP）。

传播节点设定为混乱地转发普通啁啾使得这些数据包向着积分函数的大方向前进。（值得注意的是，传播节点可以同时被专用和公用数据流使用——像之前一样为所有信息提供传输。）

积分函数可以被配置用来从它发现的相关终端设备上收集数据，这些设备是它通过寻找源自特定类别的设备、位置或其他特性的小数据流发现的。这些积分函数会合并这些源自被任何数量的未知个体安装的许多独立终端设备的数据流，去形成一个有趣的大数据信息。

13.3.5 编程和"偏好"

积分函数的人工编程可以指导它通过传统互联网去寻找特定地点和类型的数据流，或者通过与已知源的亲和程度，积分函数能够识别潜在的相关候选数据流。在互联网上定位到适当的湿度传感器数据流后，积分函数开始接收并整合这些数据。积分函数甚至可能"偏向于"带有传播节点的发布代理，为的是在小的数据流中把啁啾合并为更大的数据包或丢弃重复的啁啾时更有效率。

积分函数的人工编程可能对这些关于湿度的数据流进行整合来寻找代表干涸程度超过阈值的变化。额外的数据，例如天气预报、空气温度和灌溉水库水位（这些信息从各种来源和资讯中获取，既有基于啁啾的也有通过互联网的），也被整合进来用以提供当前和未来一段时间完整的灌溉需求图。

由此产生的报告可能会提供给用户作为行动参考。或者，在更为自动化的情形里，积分函数（通过对它的编程）会进行响应来改变某一块地的浇水次数或者持续时间（如果灌溉阀门也是由物联网控制的话）。在这个应用中，积分函数可能还会分析视频监控数据流来确认喷灌是否打开和正常运转。

值得注意的是，这种功能物联网网络可以和任意的传输拓扑结构所交互。农业企业无需为全部的传输路径建立自己专用的网络；相反，它可以使用传统互联网作为其传输数据的基本构架。企业可能只需要适当地配置湿度传感器和一些专门的传播节点即可。

这个例子只是数以百万的物联网应用的一个缩影。但是，包括及其简单的终端设备、发布和订阅、利用公共网络传输和各种数据源的整合，这些概念都被广泛地应用在了物联网中。

13.3.6 面向接收者的选择性

和雌性植物只"选择"同种植物并排斥其他植物的花粉或者灰尘等一样，积分函数在选择哪些啁啾数据流作为输入去分析合并的时候也具有选择性。

积分函数可能被编程来"设定"、配置或者操纵终端设备，这是通过产生它们自己的"啁啾"数据流，而该数据流被压缩以通过传统的互联网到达一个接近目标终端设备的已知传播节点。借助于设定终端设备方向的传输"箭头"，这些数据包被传递到合适的传播节点（通常是以 IPv6 的形式），之后以啁啾数据包的形式输出。积分函数可能将多个分散终端设备的啁啾整合到一个广播数据包内，在需要的情况下这个广播数据包会被中继传播节点修剪然后再次广播出去。

终端设备可能能够"监听"到大量的数据流，但是由于相似的面向接收者的选择性，这些终端设备只对面向它们的特定数据流产生动作。如前所述，中继路由和寻址信息是传播节点的主要功能；终端设备只需要检测简单的物联网啁啾地址。

第 14 章　云计算

云计算为计算资源的供给提供了新的模式。这种模式把计算资源的位置转移到了网络，以减少与硬件和软件资源相关的管理成本。它代表了长久以来将计算作为一种实用工具的愿景，也就是规模化的经济原理有助于有效地降低计算资源的成本。云计算简化了硬件配置、采购和软件部署的耗时过程。因此，对于数据密集型应用的部署来说它具有许多优势，例如：如资源的弹性，付费使用的成本模型，低的上市时间，以及对无限资源和无限扩展性的认知。因此，如果工作量增加，通过连续加入的计算资源，实现无限的吞吐量至少理论上是有可能的。

要充分利用云托管数据存储系统，很好地了解云计算技术的各个方面是非常重要的。本章从如下方面对云计算进行概述：关键定义（第 14.1 节），相关技术（第 14.2 节），服务模式（第 14.3 节）和部署模型（第 14.4 节），之后第 14.5 节分析了当前最先进的公共云计算平台，其中该平台专注于自己的置备能力。第 14.6 节总结了构建使用云计算技术软件应用的商业利益。

14.1　定义

云计算是一种导致计算进化的新兴趋势，其建立在几十年间对虚拟化、自主计算、网格计算和效用计算、以及最近对网络和软件服务技术的研究之上。尽管云计算现在已经被广泛接受，但云计算的定义一直有争议，这是由于构成云计算的整体视图技术的多样性导致的。从研究的角度来看，许多研究人员已通过延伸自己的研究领域范围提出了他们对云计算的定义。从面向服务架构的观点，杜布罗夫尼克认为云计算是"*面向服务的体系结构，它包括降低了最终用户的信息技术开销，具有更大的灵活性，降低了总体拥有成本，按需服务，以及许多其他的东西*"。Buyya 等从集群和网格的角度得出云计算的定义，并称赞服务供应商和客户之间服务级别协议（SLA）的重要性，将云计算描述为"*一类并行和分布式系统，它由相互连接和虚拟化的计算机集合构成，其中这些计算机是动态供应的，并作为基于 SLA 的一个或多个统一的计算资源*"。安布拉斯特等从 Berkeley 强调云计算的三个方面，包括按需提供的无限计算资源的错觉，没有前期的承诺和付费使用的工具模型，将云计算称为"*包括通过互联网、数据中心硬件和系统软件的服务应用程序*"。此外，从行业的角度来看，更多由行业专家做出的定义和摘录，从可扩展性、弹性、商业模式和其他角度对云计算进行分类。由于存在大量由不同技术引起的怀疑和混乱，同时营销炒作盛行，导致很难达成一个关于云计算定义的一致协议。出于这个原因，美国国家标准与技术研究所一直致力于研究提出云计算的一个准则。准则中云计算的定义已经得到广泛的认可。它被描述为：

"*云计算是一个模型，它能够方便地按需访问到网络可配置的计算资源共享池（如网络，服务器，存储，应用程序和服务），而这些资源能够以最小的管理工作或服务提供商交互来实现快速配置和发布*"

根据这个定义，云计算具有以下的基本特征：

1. *按需自助服务*。消费者可以根据需要自动地单方面提供计算能力，例如服务器时间和网络存储，而无需与每个服务的提供者进行人工交互。

2. *广泛的网络访问*。云计算的功能被提供在整个网络上，并且可以通过标准机制进行访问，该机制促进了不同客户端平台的使用（例如，移动电话，笔记本电脑和 PDA）。
3. *资源池*。提供者的计算资源被集中，借助多租户模式，通过根据消费者的需求动态分配和重新分配不同的物理和虚拟资源来服务多个消费者。存在位置无关的意识，即用户通常对所提供的资源的准确位置没有控制权或者了解，但或许能够在更高的抽象级别上指定位置（例如，国家，州，或数据中心）。资源包括存储，处理，存储器，网络带宽，虚拟网络和虚拟机。
4. *快速伸缩*。云计算在某些情况下，具有自动迅速且弹性调配的能力，从而能够快速地向外延展并迅速释放实现快速伸缩。对消费者来说，这种能够调配的能力似乎经常是不受限制的，能够以任意数量随时购买到。
5. *测量服务*。云系统通过利用其针对某类服务抽象测量能力，自动地控制并优化资源使用（例如，存储，处理，带宽，和活动的用户帐户）。资源的使用可以被监视、控制和报告，这样一来提供了对利用服务的提供者和消费者的透明度。

14.2 云计算的相关技术

云计算是从过去几十年间对不同相关技术的研究中演变出来的，从中它也继承了一些特性和功能，如虚拟化环境、自主计算、网格计算和效用计算。图 14.1 显示了对云计算在托管软件应用中的演变。事实上，云计算经常被拿来与以下技术进行比较，它们每一个都与云计算有着一些共通之处。表 14.1 对这些技术和云计算之间功能差异进行了简明的总结，而相关技术的细节则论述如下：

虚拟化

虚拟化是为高层应用而隔离和抽象了低层资源，并提供虚拟化资源的技术。在硬件虚拟化环境中，在虚拟机管理程序支持下物理硬件的细节可以抽象出来，如基于 Linux 内核的虚拟机和 Xen。由管理程序管理的虚拟化服务器通常被称为虚拟机。在一般情况下，多个虚拟机可以从单个物理机器被抽象出来。随着物理机的集群，虚拟机管理程序能够提取和集中资源，并根据需要动态分配或重新分配资源到虚拟机上。因此，虚拟化形成云计算的基础。由于虚拟机是和底层硬件与其他虚拟机相分离的。供应商可以通过提供运行在作为服务器的虚拟机中的应用程序或者直接访问虚拟主机的方式，通过自定义平台来满足客户的需求，这样一来用户将被允许用他们自己的应用构建服务。此外，云计算不仅是对资源的虚拟化，也是出于管理客户竞争资源需求的考虑，对资源进行智能分配。图 14.2 示出虚拟化技术在云计算环境的使用范例。

自主计算着眼于建设一个自我管理的计算系统，这意味着其能够在规定的一般政策和规则进行操作，而无需人工干预。自主计算的目的是为了克服迅速增长的计算机系统管理的复杂性，同时能够保持增加互联性和集成度不减。虽然云计算在互联和集成各大洲的分布式数据中心方面，表现出与自动计算具有一定的相似性，但其目的在某种程度上是降低资源成本，而不是降低系统的复杂性。

网格计算

网格计算是一种分布式计算模式，它通过协调网络资源，实现一个共同的计算目的。网格计算的发展最初是由通常是计算密集型的科学应用所推动，但需要大量数据传输与处理的应用程序也能够利用网格的优势。云计算似乎与网格计算是相似的，因为它也是采用分布式资源来实现应用程序级的目标。然而，云计算需要更进一步通过利用虚拟化技术来实现按需资源共享和动态资源供应。

效用计算

效用计算代表了将资源包装为类似于传统公用事业公司提供的计量服务的商业模式。尤其是，它允许根据需求和收费进行资源配置，这一点是基于用途而不是均一费率的。效用计算的主要好处是更好的经济性。

云计算可以被视为一种效用计算的实现。随着按需资源供应和公用事业为基础的定价，用户能够获得更多的资源来处理意料之外的高峰，而只支付他们所需的资源；同时，服务供应商可以最大限度地提高资源利用率和降低运营成本。

14.3 云服务模型

准则中定义了如今被广泛接受的三种云服务模型的分类。这三个服务模型，即基础设施即服务（IaaS），平台即服务（PaaS）和软件即服务（SaaS）。如图14.3，三个服务模型构成了云计算堆栈结构，分别是软件即服务在顶部，平台即服务在中部，基础架构即服务在底部。而倒三角的结构显示出每种模型供应商的可能比例，值得一提的是，准则中三个服务模型的定义多是从用户的角度出发的。与此相反，巴克罗等从供应商的角度定义了三种服务的模型。

1. *基础架构即服务*：通过虚拟化，提供者能够分裂、分配和动态调整云资源，包括加工、存储、网络，以及其他基本的计算资源来根据客户要求构建虚拟化系统。因此，客户能够部署和运行任意操作系统和应用程序。客户并不需要部署底层的云基础构架，但可以控制需要部署哪些操作系统、存储选项和已经部署的应用程序，但是对于网络组件的选择只有有限的控制权。典型的提供商是 Amazon Elastic Compute Cloud（EC2）和 GoGrid。

2. *平台即服务*：提供商提供了一个额外的抽象级别，这是一个系统在其上运行的软件平台。包括网络，服务器，操作系统，或存储的云资源的变化是以一种透明的方式进行的。用户不需要对云资源进行部署，但对已经部署的应用和可能的应用托管环境配置具有控制权。该领域知名的三个平台分别是谷歌的 App Engine，微软 Windows Azure 平台，和建立在 Amazon EC2 之上的 Heroku 平台。第一个平台提供 Python，Java 和 Go 作为其编程语言。第二个支持.NET Framework，Java，PHP，Python 和 Node.js 语言。而第三个平台兼容 Ruby，Node.js，Clojure，Java，Python 和 Scala 语言。

3. *软件即服务*：提供商为多种托管在其云基础架构的客户提供了带有潜在利益的服务。该服务可以通过瘦客户端接口，如网页浏览器，被各种客户端设备所访问。用户不需要管理云资源甚至单个应用程序的功能。可能的话，用户可以被授予有限的用户专用应用配置设置。大量软件及服务的提供商可以在互联网上找到，其中包括 Salesforce.com，Google Apps，and Zoho。

14.4 云部署模型

该准则还定义了四种类型的云部署模型，其描述如下：

1. *私有云*。一个机构的专用云。它可能由该组织或第三方进行管理，可能存在开通或关闭的前提。私有云提供了最高程度的对性能、可靠性和安全性的控制。然而，他们经常被批评为类似于传统的专有服务器群并没有提供诸如没有前期资本成本的效益。

2. *社区云*。该云基础构架是由一些组织共享，并支持有共同关注的问题的特定社区（例如，出于使命、安全要求、政策和遵守的考虑）。

3. *公共云*。该云基础架构是提供给广大公众或大型产业集团的，由一个销售云服务的机构（如亚马逊，谷歌，微软）所拥有。由于云服务客户的需求不同，服务供应商必须确保他们能够灵活的提供服务。因此，使用服务水平协议（SLA）来规定公共云提供的服务的质量，它代表提供商和用户之间的关于规定用户需求和提供商承诺的合同。通常，一个 SLA 包括的项目如正常运行时间、隐私、安全和备份程序。在实践中，公有云为服务用户提供了几个关键的利益，如：包括对基础构架没有初始资本投资和将风险转移到基础构架提供商。然而，公共云缺乏对数据、网络和安全设置的细粒度控制，这可能妨碍其在许多业务场景的效力。

4. *混合云*。该云基础构架是两个或多个云（私有的、社区的、或公共的）组成的，而它们仍是唯一的实体，通过标准化或使数据和应用程序具有可移植性的专有技术结合在一起（例如，使云之间负载均衡的云爆发）。特别是，云爆发是一种用于混合云，根据需要为私有云提供额外资源的技术。如果私有云具有处理其工作负荷的能力，则不需要使用混合云。当负载超过私有云的能力，混合云自动向私有云分配更多的资源。因此，混合云提供了比公共云和私有云更大的灵活性。具体来说，它们比起公共云提供了对应用程序数据更严格的控制和安全性，同时还便于按需服务膨胀和收缩。不利的一面是，设计一个混合云需要仔细确定公共云和私有云组件之间的最佳分割。表 14.2 总结了在所有权、用户权、位置和安全性方面四个云部署的模型。

14.5 公共云平台：体现最高水平

在公共云计算领域的主要参与者包括 Amazon Web Services、Microsoft Windows Azure、Google App Engine、Eucalyptus 和 GoGrid，它们提供各种用于监控、管理和调配资源预包装服务。然而，在这里每个云所采用的技术确实不尽相同。

对于 Amazon EC2，有三个 Amazon 服务，即 Amazon Elastic Load Balancer、Amazon Auto Scaling 和 Amazon CloudWatch，它们共同构成其在 EC2 上所需承担的应用服务配置的功能。Elastic Load Balancer 服务自动规定传入的应用工作负载可用于 EC2 实例，而 Auto Scaling 服务可用于动态缩放和扩展 EC2 实例的数量，用以处理服务需求模式的变化。最后的 CloudWatch 服务可以与基于收集实时信息的战略决策的上述服务进行整合。

Eucalyptus 是一个开源的云计算平台。它包括三个控制器。在这些控制器中，群集控制器是支持应用服务提供和负载平衡的一个关键组成部分。每个群集控制器托管于群集头部节点，用以与外部公网和内部私网进行交互。通过监控在服务器控制器的池实例的状态信息，群集控制器可以选择任何可用的服务/服务器来配置传入请求。然而，相比于 Amazon 服务，Eucalyptus 还缺少一些关键功能，如自动缩放其内置的置备程序。

从根本上说，Microsoft Windows Azure fabric 具有编织状结构，它由包括服务器和负载平衡器以及包含电源和以太网的边缘的节点构成。fabric 控制器通过一个名为 Azure Fabric Controller Agent 的内置服务来管理服务节点，该内置服务在后台运行，同时跟踪服务器的状态，并将这些指标报告给控制器。如果有故障状态报告，该控制器可以重启服务器或者将服务从当前的服务器向其他正常服务器迁移。此外，借助匹配符合要求的需求的虚拟机，该控制器还支持服务供应。GoGrid 的云主机为开发人员提供了 F5 负载均衡器，目的是分配跨服务器的应用业务流，前提是这些服务器的 IP 地址需要和特定端口连接。负载均衡器提供了路由应用服务请求的轮循算法和最少连接算法。此外，负载平衡器能够检测到服务器崩溃的发生，进一步重定向请求到其他可用的服务器。但目前，GoGrid 只为开发人员提供了一套可编程的 API 来实现他们自定义的自动缩放服务。

不像其他的云平台，Google App Engine 为开发人员提供了一个可扩展的平台，在这个平台上应用程序可以运行，而不是提供对一个定制虚拟机的直接访问。因此，在 App Engine 中对底层操作系统的访问是受限的，其中的负载均衡策略、服务提供以及自动缩放全部由系统进行管理，其具体的实现过程是未知的。Chohan 等已经提出了在 Amazon EC2 和 Eucalyptus 顶部建立类似 App Engine 框架的 AppScale 的初步努力。他们的成果包括了自动部署、管理、缩放和容错的 App Engine 应用程序的多个组件。在他们的设计和实施中，单个 AppLoadBalancer 存在于 AppScale，目的是分配用户对 App Engine 应用程序中 AppServers 的初始请求。用户最初接触 AppLoaderBalancer 是请求登录到 App Engine 应用程序。之后，该 AppLoadBalander 验证登录并将请求重定向到一个随机选择的 AppServer。一旦请求被重定向，用户可以开始与 AppServer 直接对话，而无需在当前会话期间经过 AppLoaderBalancer。位于 AppLoadBalancer 中的 AppController 还负责监视 AppScale 部署时 AppServers 的增长和收缩。

没有单个的云基础构架提供商在整个世界上的所有可能的地点都有自己的数据中心。这样一来，所有的云应用提供商目前对他们所有的客户都很难满足 SLA 期望。因此，这是合乎逻辑的：每一个提供商将建立定制的 SLA 管理工具来提供更好的支持，以满足其特定需求。这种需求往往产生于全球运营和应用的企业，如互联网服务，媒体托管和 Web2.0 应用程序。这就需要为云基础架构服务提供商的无缝集成建立技术和算法，目的是在不同的云服务提供商的服务间进行配置。

14.6 云计算的商业效益

随着云计算的发展，企业会占用共享的计算和存储资源，而不是建设、运营以及改善自己的基础架构。市场变化的速度对企业 IT 基础架构的适应和交付产生了显著的压力。原则上，云计算使企业能够获得灵活且经济高效的 IT 基础架构，类似于国家电网使家庭和企业能够接入到一个集中管理、高效和经济有效的能源来源。当从自己供电的局面中解脱出来，企业能够专注于自己业务的核心竞争力和客户的需求。特别是，云计算技术提供了用于构建软件应用的一些明显的商业利益。

这些利益的实例是：

1. *没有前期基础构架投资*：建设一个大型系统可能会花大价钱投资于不动产、硬件（架，机器，路由器，备用电源）、硬件管理（电源管理，散热系统）和操作人员。由于高昂的前期费用，甚至在项目开始之前，通常就需要经过几轮的管理审批。而有了云计算，不需要固定成本或者启动成本就可以开始你的项目。

2. *适时的基础架构*：在过去，如果您的系统颇有名气，但同时其基础构架不能很好地扩展，那么您的应用程序可能会成为其成功的牺牲品。在另一方面，如果您投入巨资而没有得到一定的知名度，您的应用程序则成为您失败的牺牲品。通过在云环境中部署应用程序，应用程序可以顺利地随着您的成长而扩展。

3. *更有效地利用资源*：系统管理员通常会担心硬件采购（当他们用完储量时）和更好的基础架构的利用率（当他们有过剩和闲置产能时）。有了云计算技术，他们可以根据现收现付的理念，通过让应用程序按需请求资源，从而更有效地管理资源。

4. *缩短处理时间的潜能*：并行是众所周知的加快处理速度的技术之一。例如，如果你有一个可以并行运行的计算密集型或数据密集型工作，在一台机器上处理需要 500 小时。采用云技术，将有可能产生并启动 500 个实例，相同的工作可以在 1 个小时内处理完成。拥有可用的弹性基础架构，使得应用程序有能力以经济有效地方式利用并行来缩短总的处理时间。

参考文献

[1] Douglas E. Comer．Computer Network and Internets．北京：清华大学出版社，1998.
[2] Douglas E. Comer．Computer Network and Internets．北京：清华大学出版社，1998.
[3] Barry Brey. Intel Microprocessors: Architecture, Programming, & Interfacing(6th Edition). Prentice Hall, 2002.
[4] J. Glenn Brookshear. Computer Science: An Overview (6th Edition). Addison Wesley Publishing Company, 1999.
[5] Rogers，D.F.著计算机图形学的算法基础（英文版·第 2 版）．北京：机械工业出版社，2002.
[6] John D Carpinelli 著．计算机系统组成与体系结构（英文版）．北京：人民邮电出版社，2001.
[7] Tanenbaum,A.S.著．结构化计算机组成（英文版·第 4 版）．北京：机械工业出版社，2002.
[8] Dhamdhere D M 著．系统程序设计和操作系统（影印版）．北京：清华大学出版社，2001.
[9] Victor P. Nelson, H. Troy Nagle, Bill D. Carroll, J. David Irwin. Digital Logic Circuit Analysis & Design．北京：清华大学出版社，1997.
[10] Lan Sommerville．Software Engineering. Addison-Wesley Publishing Company, 1992.
[11] David A. Patterson,John L. Hennessy. Computer Organization and Design—The software/hardware interface (5th Edition). Morgan Kaufmann publications, 2014.
[12] J. Glenn Brookshear, David T. Smith, Dennis Brylow. COMPUTER SCIENCE--AN OVERVIEW(11THEdition). Addison-Wesley Publishing Company, 2012.
[13] LiangZhao, Sherif Sakr, Anna Liu, Athman Bouguettaya. Data Clound Management, 2014.
[14] Francis daCosta. Anatomy of the Internet of Things. Rethinking the Internet of Things, pp 23-40,2014.
[15] 王秉钧，郭正行编著．科技英汉汉英翻译技巧．北京：天津大学出版社，1999.
[16] 张淑方编著．科技英语捷径．北京：北京语言学院出版社，1990.
[17] 刘兆毓主编．计算机英语．第 2 版．北京：清华大学出版社，1997.
[18] Tang Renyuan, Sun Jianzhgong, etal．Optimization of Electromagnetic Devices Using Intelligent Simulated Annealing Algorithm．*IEEE Trans. on Magn*. 1998, (5):2992-2995.
[19] 邵启祥著．科技英语翻译中的陷阱·误区及其他．北京：国防工业出版社，1991.